Fundamental Aspects of Pollution Control and Environmental Science 4

POLLUTION IN HORTICULTURE

Fundamental Aspects of Pollution Control and Environmental Science

Edited by R.J. WAKEMAN

Department of Chemical Engineering,
University of Exeter (Great Britain)

1
D. PURVES
Trace-Element Contamination of the Environment

2
R.K. DART and R.J. STRETTON
Microbiological Aspects of Pollution Control

3
D.E. JAMES, H.M.A. JANSEN and J.B. OPSCHOOR
Economic Approaches to Environmental Problems

4
D.P. ORMROD
Pollution and Horticulture

Other titles in this series (in preparation):

S.U. KHAN
Pesticides in the Soil Environment

R.E. RIPLEY and R.E. REDMANN
Energy Exchange in Ecosystems

W.L. SHORT
Flue Gas Desulfurization

A.A. SIDDIQI and F.L. WORLEY, Jr.
Air Pollution Measurements and Monitoring

D.B. WILSON
Infiltration of Solutes into Groundwater

Overleaf: The use of indicator plants for the
detection of air pollution

Fundamental Aspects of Pollution Control and Environmental Science 4

POLLUTION IN HORTICULTURE

D.P. ORMROD

*University of Quelph, Ontario Agricultural College,
Department of Horticultural Science,
Quelph, Ontario (Canada)*

ELSEVIER SCIENTIFIC PUBLISHING COMPANY

Amsterdam — Oxford — New York   1978

ELSEVIER SCIENTIFIC PUBLISHING COMPANY
335 Jan van Galenstraat
P.O. Box 211, 1000 AE Amsterdam, The Netherlands

Distributors for the United States and Canada:

ELSEVIER NORTH-HOLLAND INC.
52, Vanderbilt Avenue
New York, N.Y. 10017

Library of Congress Cataloging in Publication Data

Ormrod, D   P
    Pollution in horticulture.

    (Fundamental aspects of pollution control and
environmental science ; 4)
    Bibliography: p.
    Includes indexes.
    1. Plants, Effect of pollution on. 2. Horticulture.
3. Pollution. I. Title. II. Series.
SB744.5.O74   632'.1   78-16598
ISBN 0-444-41726-5

ISBN 0-444-41726-5 (Vol.4)
ISBN 0-444-41611-0 (Series)

Printed in The Netherlands

PREFACE

This book is intended to provide a broad scientific assessment of the interaction of horticulture with pollution. For convenience, pollutants are discussed under the headings air, trace elements, pesticides, water and salt. The book is intended for the use of horticultural scientists involved in research, education and extension. Information is presented at the senior undergraduate or beginning graduate student level. Extensive use is made of journal references and responses are frequently illustrated with tables, figures and photographs taken from scientific literature.

Generally, for any particular species, the information available on pollutant sensitivity is incomplete. This is all the more problematic because species responses differ widely within the same family, and even cultivar responses frequently differ within species. The results of air pollution research on a large number of plant species appear in a wide range of pollution-oriented journals, many of which are not commonly reviewed by horticulturists. While some vegetation studies reported in such journals are not of horticultural species all such reports may have some relevance to horticultural problems, and research planning should involve the knowledge of all plant responses regardless of species or publication medium. Extensive use has been made of less readily available journal articles as well as those in the horticultural journals. An exhaustive review of literature is beyond the scope of this presentation, particularly for the non-air pollutants. Instead the relevance of all these pollutants to horticulture is introduced and representative examples are provided.

The format of the book is to present in Part I information on the characteristics of the pollutants in a particular group, i.e. air, trace elements, pesticides, water or salt, and their impacts on plant physiology, biochemistry and genetics; then to present in Part II a discussion of the available knowledge of horticultural plant responses to each major pollutant in each group. Species are considered separately, wherever possible, because knowledge of individual species and cultivar responses is the ultimate need of the horticultural scientist. Indexed appendices of chemical names and species and a glossary of pollution terms are provided. The reference list is also indexed and there is a general index.

An over all view of the interactions between horticulture and pollution has been provided by Day (1970). His concepts readily form the basis for a thorough consideration of pollution in horticulture. The message is that the relationship between horticultural practices and pollutants is incredibly complex. It is the purpose of this book to unravel as many as possible of these complexities to form the basis for devising practices which will minimize the impact of pollutants on horticulture.

This book is dedicated to L.H. Wullstein whose enthusiasm and interest provided the initial impetus to my pollution studies and to N.O. Adedipe who energetically conducted the first pollution experiments in our laboratory. I gratefully acknowledge the assistance in the pollution research programme during the past decade of students, technicians and other colleagues almost too numerous to mention, including R. Baker, D. Ballantyne, R. Barrett, T. Blom, M. Czuba, T. Elkiey, D. Evans, W. Evans, R.A. Fletcher, J. Gardner, E. Heale, G. Hofstra, A. Humphreys, H. Khatamian, B. Lagassicke, G. Lumis, L. Neil, D. Papple, M. Phillips, J. Psutka, L. Pyear, C. Rajput, G. Rebick, G. Stephenson and M.J. Tsujita. Many thanks as well to M.E. Pyear for typing the manuscript and to E. Ormrod for many hours spent in proof-reading. I am also grateful to J. Proctor, R. Reinert and T. Tibbitts for reviewing the first draft of the manuscript.

-- D.P. Ormrod
Guelph, Ontario, Canada 1978.

CONTENTS

INTRODUCTION

Pollution interacts strongly with horticulture. Much intensive horticultural activity is concentrated in or near urban areas where pollutant concentrations tend to be high as a result of transportation systems, space heating, industrial activity, and other human-oriented activities. Also, the intensive nature of horticultural production creates its own pollution problems arising from the need for frequent heavy use of pesticides and fertilizers. These circumstances require horticulturists to have a better understanding of pollution problems than any other members of the agriculture and food system.

The role of horticulture in providing attractive, safe and nutritious fruit and vegetable foods is beset with pollution problems. Air pollutants such as ozone (O_3), sulphur dioxide (SO_2), fluoride (F) and other forms of pollution may visibly injure plants, impairing consumer acceptance, and reducing growth rates and yield. There are potential health hazards associated with the consumption of foods high in some elements and pesticides. Again, horticulture is particularly vulnerable to such problems. Fruit and vegetables produced in the vicinity of metal processing industries or major highways may accumulate trace elements associated with the industries or with traffic.

The role of horticulture in providing an aesthetically pleasing landscape is also hard pressed by the existence of numerous pollutants which cause the development of visible injury symptoms on landscape and turf plants and flower crops. Growth and development rates may be impaired in ornamental species as well as in fruit and vegetable crops. Frequently, the injury is greatest in the very locations in which horticultural beautification of the landscape is most needed.

Ideally pollutants should be abated at their sources but the planning and executing of abatement programmes are the roles of chemists and engineers not horticulturists. The role of the horticulturist is to identify species which are insensitive of pollutants, to devise management systems which minimize the impact of pollutants on horticultural plants, and to search for protectants against pollutants. Fortunately there are relatively insensitive cultivars within each species suggesting that plant breeding and selection programmes may be very successful. Also it is fortunate that there are large differences in the effects of management practices on pollution responses. As well, numerous protectants have been shown to be effective against various pollutants.

There are occasionally beneficial effects of agents normally injurious. For example, SO_2 may in some cases meet nutritional needs for sulphur. Such desirable effects of some pollutants under some circumstances are a complication in our general attitude that all pollutants are harmful and should be eliminated if possible.

One purpose of this book is to provide the horticulturist with an understanding of the chemical nature, occurrence and importance of air, trace element, pesticide, water and salt pollution. Another purpose is to present an overview of the effect of these pollutants on individual horticultural species or groups of related species. Emphasis is placed on species and cultivar responses, on relationships between these responses, management practices, and the environment, and on methods of ameliorating the harmful responses. Finally, some thoughts are shared on the future relationship between pollution and horticulture. The horticulturist in the modern world should be familiar with the undesirable impacts of man's activities on fruit, vegetables and ornamental plants. Only with knowledge will the horticulturist recognize pollution-oriented problems and take adaptive and protective action to maintain plant quality. It is the purpose of this book to provide knowledge of the problems and an awareness of their solutions. The interaction of pollution and horticulture thus provides both challenges and opportunities for the horticulturist.

PART I

POLLUTANTS AND THEIR EFFECTS

CHAPTER 1

AIR POLLUTION

1.1 Introduction

Air pollution may reduce growth and yield of horticultural crops with the reduction dependent upon the pollutant, its concentration, the duration and timing of exposure, the sensitivity of the plant, environmental conditions, and other interacting factors. Pollution may also mark and discolour foliage, reducing the visual attractiveness and therefore the value and marketability of fruits, vegetables, and ornamental plants.

Much variation in plant injury may be noted among plants grown in different environments. This represents not only variation in pollutant levels but also differential sensitivity to climatic factors such as temperature, light and humidity and to edaphic factors such as nutritional and water status, which in turn affect sensitivity. Differing sensitivities are also observed among leaves on the same plant; for example, immature leaves may be relatively insensitive to air pollutants whereas recently mature leaves are usually most sensitive. Additional variability can be noted among leaves of approximately the same age. The leaves are the most sensitive organs of the plant and generally display more visible injury than flowers or fruit.

Research concerning urban and industrial air pollution on plants has increased greatly in recent years. Most of the early studies involving economic plants were on sensitive field crops such as tobacco and white beans. Many of the recent research reports have focused on horticultural plants because of concern for yield and appearance. Many vegetable, fruit and ornamental species have been found to be sensitive to one or more commonly occurring air pollutants singly or in combination. There are marked differences in sensitivities to most air pollutants, both between and within species. This has led to the suggestion that plant breeding should aim at developing insensitive plants. Only tolerant cultivars would be planted in areas where air pollutants were expected to be present. Unfortunately this has the obvious disadvantage of limiting consumer aesthetic as well as intrinsic choices. There is a wide diversity of plant response to pollutants and this combined with the large number of economically important horticultural species indicates that much remains to be done in horticultural research involving air pollution.

1.2 Literature Reviews

There have been many reviews of the scientific literature on plant res-ponses to pollutants. Many of these reviews have provided extensive information on horticultural plant responses. An early illustrated review of plant injury due to atmospheric pollution had examples largely of horticultural species (Middleton et al. 1958). Another illustrated guide to identification of air pollution injury on many horticultural and other crops, resulting from exposure to many different gases, included suggestions for the use of indicator plants for detecting the kind and amount of air pollutants but cautioned that such res-ponses must be distinguished from those caused by other plant stresses (Darley et al. 1966).

Useful reviews include one on symptoms, factors affecting response, field monitoring, response evaluation, and methods of reducing injury (Brandt and Heck 1968); another discusses the role of air pollutants in shifting populations from the view of plant pathology with examples drawn largely from plant pathogen-air pollutant interaction studies (Treshow 1968). There are also concise but thorough reviews of air pollutant-incited disorders (Heggestad 1968; Wood 1968; Daines 1969a). Considerable information on air pollution is included in a book on environmental factor effects on plants (Treshow 1970a).

Available information up to 1970 concerning air pollution injury to plants is compiled in a pictorial atlas (Jacobson and Hill 1970). Another very thorough review includes identification of the major pollutants, symptoms of injury, sources and mixtures, growth effects, time-concentration-injury relations, inter-acting factors, and use of tolerant cultivars and species (Heggestad and Heck 1971). A general review of the impact of air pollutants on plants is available (Waggoner 1971) as well as reviews of the relationships of pollutants to agricul-tural production with discussion of the pollutants generated by agriculture (Heggestad 1974; Rich 1975). An illustrated manual for the diagnosis of air pollution injury has been written by Skelly and Lambe (1974). A survey of air pollutant effects on forest and tree species includes specific examples of cause and effect relationships (Hepting 1968).

A number of the articles of relevance to horticulture are included in a volume consisting mainly of review papers presented at an air pollution seminar (Mansfield 1976). A book has been published which includes an extensive survey of the literature on the effects of SO_2, HCl and HF on vegetation together with the results of many of the author's own experiments (Guderian 1977). The

evidence for effects of long-term low levels of air pollution on plant growth
and development has been compiled (Feder 1973). Examples are given to illustrate
that plants may have depressed growth and yield without visible injury symptoms.
Research is called for on the effects of long exposures to low levels of pollutants and on the relationship between growth effects and visible injury. Air
pollutants have been treated as stress factors on food plants and particular
attention has been given to apparent yield effects (Howell 1974). Emphasis is
given to the need for determination of the interaction of pollutants with other
stress factors including drought, temperature, wind, frost, hail, pesticides,
pathogens, nutritional deficiency, weeds, and soil pH. The effects of air pollution on plants in relation to meteorological aspects and instrumentation have
also been reviewed (Mukammel 1976). Air pollutants are considered as a part of
the broader context of polluting chemicals in the environment in another review
(Ten Houten 1973).

There are numerous reports of air pollution problems in particular regions.
For example, the diversity of plant injury caused by pollution in New Jersey,
U.S.A., has been described including apparent effects of SO_2, O_3, PAN and
possibly aldehyde (Daines et al. 1960a; Brennan et al. 1967). A wide range of
injury symptoms is evaluated and O_3 is apparently the principal phytotoxic air
pollutant. A number of sensitive horticultural species are identified and year-
to-year and month-to-month variation in injury is related to sources and meteorological effects (Table 1). Another summary of field observations has been

Table 1. Monthly frequency of plant damage caused by various pollutants from
1961-1966 in New Jersey, U.S.A.

Month	SO_2	O_3	PAN
March	3	3	0
April	5	4	1
May	9	11	1
June	3	12	3
July	1	14	0
August	0	9	2
September	0	3	0
October	0	5	1
November	0	1	1

Data from Brennan et al. (1967)

reported which includes descriptions of plant injury due to SO_2, HF and Cl_2
(Hindawi 1968). Typical sites are New York City, U.S.A. for SO_2 injury; near

a glass fiber plant for F injury; and near a manufacturing plant for Cl_2 and HCl mist injury. Descriptions of visual symptoms are supported by chemical and microscopic analyses. The effect of atmospheric dispersion of pollutant plumes on plant exposure has been described (Smith 1968). It is suggested that experimental exposures be based on field measurements of actual concentration variations to which plants may be exposed.

Horticultural Reviews. A few literature reviews are directed specifically to horticultural crop responses. In one review the major gaseous pollutants and their effects on many horticultural species and cultivars are described with tabulated references to air pollutant effects on growth and yield and a number of conclusions about growth and reproduction effects (Reinert 1975). A discussion of needs for research on air pollution effects on horticultural plants is also provided. A broad review of horticultural research in air pollution includes consideration of species and cultivar responses, metabolic effects, environmental influences, and protection methods (Ormrod et al. 1976). Glasshouse air pollution problems are also described (Hand 1972). Extensive data on landscape plant sensitivity to air pollution have been compiled for landscape horticulturists (Odoi 1976). Tables are also provided of sources, injury symptoms and effects on leaves of differing ages, of a wide range of pollutants.

The place of horticulture within the broader context of environmental quality has been considered with emphasis on the impact of pesticides and nutrients along with some consideration of crop residues, salinity, and erosion sediment problems (Walker 1970).

1.3 Economic Losses

The problems and principles of estimating pollution damage are outlined by Saunders (1976). Examples are provided for a wide range of species and many kinds of pollutants. Losses sustained by horticultural producers as a result of pollution are difficult to assess due to a lack of precise information on how air pollutants affect the marketability of the crop. It is important to have valid assessments of horticultural losses due to pollutants to help guide pollution control efforts and to set practical environmental quality standards. Our present information on air pollution effects is too incomplete to allow prediction of total impact on horticultural production in terms of yield and quality reduction (Heck 1973), although a number of sensitive problem species have been identified. Long-term studies are needed to assess accurately the impact on production of many crops. The nature of such studies has been described with the presentation of a model for determining the effect of air pollution on agriculture, consisting of experimental determination of pollutant gases, observation of environment, pollutant and plant development interactions, and interpretation of the economic and social impacts of the changes (McCune 1973).

A detailed study of the economic impact of air pollutants on all plants in the U.S.A. has been conducted (Benedict et al. 1971). Losses are estimated separately for oxidants, SO_2 and fluorides. Using a formula of crop value X crop sensitivity X pollution potential, the total annual dollar loss to all crop plants in the U.S.A. was calculated as $85.5 million, of which $78 million was due to oxidants, $3.25 million to SO_2 and $4.25 million to fluorides. The corresponding figures for ornamental plants were $46, 43, 3 and 0.175 million, respectively.

The year to year variation in crop loss due to air pollutants is discussed by Pell and Brennan (1975). Pollutant concentrations may be similar between years but environmental factors different. Differences between years in rainfall, temperature, humidity, nutrition, presence of diseases, and the use of systemic fungicides may result in many-fold differences in damage.

1.4 Expressing Species and Cultivar Sensitivity

Plants often react to a particular pollutant by developing typical symptoms regardless of species, but there are exceptions in the case of every pollutant. Some pollutants, such as PAN, have a wide range of injury symptoms among species (Jacobson and Hill 1970). A number of terms employed to describe and discuss pollutant responses have somewhat unique meanings in air pollution usage. For example, precise use of the terms injury and damage has become almost universal. Injury is considered to include all abnormal plant responses resulting from the pollutant exposure including metabolic disturbance, leaf chlorosis and necrosis, and retarded growth and development. Damage is generally considered to be effects on the economic value of the plants for their desired use, such as reduced yield or decreased consumer acceptance which determines market value.

Another concept is concerned with exposure concentration and symptoms. The two general categories of exposure conditions and resultant symptoms are referred to as acute and chronic. Acute exposures are exposures to high concentrations of gases for a relatively short time resulting in rapid development of visible injury symptoms. Chronic exposures consist of long duration of exposures to low concentrations resulting in injury only after an extended period. The symptoms may differ, with acute exposure often resulting in drastic well-defined necrotic areas on leaves, leaf abscission and severe dwarfing, while chronic exposures may result only in mild chlorosis, yellowing, stippling, or subtle growth reductions. A third category may be added -- long-term effects due to the impact of pollutants over decades. The chronic and acute effects are considered to be due to the direct action of the pollutants while long-term effects may be caused indirectly, for instance, by by-products, rather than by the pollutant itself. Accumulation of trace elements in soils and vegetation over a period of years is classified as a long-term effect.

Another concept is that of visible and hidden injury. In addition to visible injury, particularly to leaves, there may be hidden injury such as growth retardation or subtle effects on plant metabolism. There may even be significant hidden injury or adverse effects upon plant growth and metabolism in the absence of visible leaf injury.

Expression of concentrations may vary among investigators. Most investigators state gaseous air pollutant concentrations as ppm, pphm, or ppb on a volume/volume basis. A few use weight per unit volume bases such as mg/m^3. The latter is subject to change with temperature and pressure following the physical gas laws, while volume/volume bases are unaffected because temperature and atmospheric pressure have the same effect on both gases.

Numerous examples of differential species and cultivar sensitivity to pollutants are presented in Part II of this book. In many cases the non-critical descriptions of injury and damage; chronic, acute and long-term exposure conditions; visible injury and hidden injury; and concentration values, have made it difficult to interpret data and draw comparisons between research reports.

1.5 Air Pollutants of Concern

Plants may be injured by one or more of a number of air polluting gases. The most troublesome on a worldwide basis are probably ozone (O_3), peroxyacetyl nitrate (PAN), nitrogen oxides (NO_x), sulphur dioxide (SO_2), fluoride (F) and ethylene (C_2H_4). The effects of these gases on plants have been extensively studied. A number of other gases may be plant-injuring pollutants in localized situations under particular circumstances. The chemistry and sources of the plant-injuring air pollutants are available in review form together with a forecast of future trends for emission of the pollutants and their precursors (Wood 1968). The predictions are for generally worsening pollutant conditions as a result of increasing population and demands for energy. The chemical basis for air pollution is also in a review which includes evidence that combustion emissions and their reaction products exert detrimental effects on crop production, increasing costs, and affecting agricultural land use planning (Schuck 1973). An outline of chemical reactions leading to injurious air pollution which considers both individual pollutants and interactions is also available (Willix 1976).

Oxidants. Oxidants in the atmosphere include O_3, organic peroxy-N compounds (such as PAN), and NO_x. Ozone and PAN are the principal plant-injuring components of a mixture called photochemical oxidant smog or simply oxidant smog. The O_3 and PAN are secondary products arising primarily from the photochemical reactions in intense sunlight of the primary pollutants, NO_x and reactive hydrocarbons. The rate of formation and maximum concentration of photochemical oxidants is a complex function of the concentration of reactive hydrocarbons, NO and NO_2 concentration and ratio, light intensity and quality, and temperature. Photochemical oxidants and their precursors move with air from their main sources in urban and industrial centers.

Photochemical oxidant pollutants have been increasing in the U.S.A., Japan, European countries, and other countries in direct proportion to increases in hydrocarbon fuel consumption. The precursors, NO_x and hydrocarbons, are emitted from both mobile and stationary sources such as internal combustion engines, industries, oil refineries, and other sources. During periods of high photochemical activity O_3 is the most important reaction product in the formation of the photochemical oxidant mixture, with O_3 accounting for 90% or more of the total oxidizing capacity. This O_3 may cause severe plant injury. PAN may also be a significant component of the mixture. Such an oxidant mixture can also be generated artificially. For example, Heck (1964) irradiated a mixture including propylene and NO_2 and obtained a gas which caused injury on leaves typical of both O_3 and PAN.

Some of the early experiments on oxidant smog effects on plants have been described (Haagen-Smit et al. 1952). Activated carbon filtration of air entering greenhouses was used to demonstrate yield depressions due to smog. The chemistry of plant-damaging smog formation was investigated by determining injuries to seven species caused by various unsaturated hydrocarbons combined with NO_x, SO_2, ultra violet light or sunlight.

In contrast to many other air pollutants, significant concentrations of O_3 may be present over large areas, sufficient to be injurious to sensitive plants, even hundreds of km away from urban and industrial centres. Hydrocarbons, such as terpenes, emitted by coniferous foliage in forested areas, and NO_x from the soil can contribute to oxidant formation.

The nature of plant injuring O_3 episodes has been described based on recording O_3 over a 5-year period (Rich et al. 1969). The daily maximum O_3 never occurs before noon and 80% of the periods of high O_3 occur in the afternoon. Daily periods of high O_3 generally last one to six hr, with an occasional daily episode lasting up to nine hr. Other studies indicate that O_3, PAN and NO_x have different daily concentration patterns.

Sensitive plants may develop symptoms of oxidant injury to leaves after two or more hr at a concentration of about 10 pphm (200 $\mu g/m^3$) or at a lower concentration over a longer time. There is wide genetic variability in response to O_3 and other oxidants and there may be subtle responses as well as visible injury. Such subtle or "hidden" responses include growth, yield, and quality reductions in the absence of visible symptoms. A number of thorough reports of oxidant effects have been presented including an investigation of the nature of O_3-induced phytotoxicity (Hill et al. 1961). Relative sensitivities of a wide range of species, including representatives of 17 families, are given along with the effects of growth stage on sensitivity. Leaf anatomical effects are described in detail, and palisade cells are noted to be most readily injured. The evidence for O_3 injury to plants in terms of both visible symptoms on the entire range of species, and hidden injury or yield effects at sub-necrotic levels is also available in review form (Richards and Taylor 1965). The problem of relating visible symptoms to O_3-induced damage is discussed in terms of crop losses. A very thorough review of O_3 and other photochemical oxidant effects on plants is now available (National Academy of Sciences 1977). Studies of growth and yield effects are particularly emphasized and several tabulations of directly comparable data are provided.

A detailed description is available of O_3 or oxidant complex injury symptoms which include pigmented lesions, upper surface bleaching, under surface glazing, bifacial necrosis, chlorosis, and chlorotic stippling or mottling (Hill et al. 1970). Species sensitivity is also rated and other disorders with similar injury symptoms noted. Among horticultural species, certain cultivars of snap bean, sweet corn, muskmelon, onion, potato, radish, spinach and tomato are relatively sensitive to O_3, as are particular cultivars or strains of carnation, chrysanthemum, grape, lilac, petunia and privet. A review of the general nature of O_3 injury to plants includes ratings of O_3 sensitivity of a number of additional species (Treshow 1970b). Among these, strawberry, aspen, bridal wreath, lilac and oak are listed as relatively sensitive with pigmented leaf senescence occurring after as little as two hr exposure to 15 pphm O_3.

A survey of plant injury in the field at Cincinnati, Ohio, U.S.A. revealed symptoms on petunia, bean, radish, squash, and tomato associated with days with high oxidant and O_3 levels (Reinert et al. 1970). The symptoms were similar to those resulting from greenhouse exposure to O_3. Many other vegetable crops and ornamental tree and shrub species in the vicinity also had stipple, fleck, and upper surface necrotic symptoms.

Several species of understorey plants in an aspen community are quite sensitive to O_3 (Harward and Treshow 1975). There are large differences among species. The weights of a few species are significantly increased in air containing up to 15 pphm O_3 while growth of almost all species is suppressed

at 30 pphm O_3. Shifts in natural community composition would thus be expected
following extended exposure to O_3.

The factors involved in the movement of O_3 molecules from the polluted air
into the plant tissue have been considered in detail (Waggoner 1971), as have
the successive stages of O_3 injury (Taylor 1973). A thorough review of the
mechanisms of O_3 entry into leaf tissue and the resultant injury syndrome is
also available (Heath 1975).

Leaf age has an important effect on sensitivity (Middleton et al. 1950).
Maximum sensitivity to controlled oxidant exposures is found in the middle-aged
leaves of tomato and the oldest leaf of bean. The relationship of leaf development to O_3 sensitivity has also been studied in soybean (Tingey et al. 1973b).
Greatest sensitivity occurs in the late stages of rapid leaf expansion when
stomatal resistance is relatively low, reducing sugars are increasing and protein,
amino acid, total soluble sugars, sucrose and starch are decreasing. Metabolite
levels remain low after injury except for total soluble sugars and reducing
sugars.

Fumigation of storage facilities with O_3 has been suggested as a method for
reducing postharvest decay of fruits and vegetables. Fungous rot of peaches and
strawberries is not reduced in $2^{o}C$ storage by 50 pphm O_3 at 90% relative humidity
but fungous growth is inhibited at $16^{o}C$ (Spalding 1966, 1968). The 50 pphm O_3
rate injures the caps of strawberries causing them to dry and shrivel. Rots of
blueberries, cantaloupes, grapes and green beans are not inhibited by O_3.
Exposed tissue of green bean fruit is injured by O_3, developing blotchy light
brown patches.

There is correlation between O_3-induced ethylene production and foliar injury (Tingey et al. 1976). The O_3-induced ethylene is produced for less than
48 hr following exposure. Of the species tested, ponderosa pine, eucalyptus and
soybean produce most ethylene in relation to O_3 concentration, while holly,
squash and marigold produce least. Ethylene production has been suggested as a
useful measure of O_3 stress.

Occasionally the plant injuring O_3 near the earth's surface is confused with
the O_3 layer at about 20 km in the stratosphere which has about 20 to 25 pphm
concentration and absorbs much of the ultraviolet radiation from the sun. The
role of this stratospheric layer in screening out excessive ultraviolet radiation is of immense importance to life on the earth's surface. Apparently little
mixing of this stratospheric O_3 with tropospheric O_3 occurs so its importance in
causing plant injury is probably minimal.

Peroxyacyl nitrates are an important group of photochemical reaction products in some areas. PAN is the most frequently found of the peroxyacyl nitrates and very low concentrations of PAN injure plants. Plant injury due to PAN has been reported occasionally from widely scattered areas. PAN injury usually appears as glazing or bronzing of the lower leaf surface but may also include tissue collapse, chlorosis, and leaf drop. Visible symptoms may be produced by as little as four ppb exposure for four hr (Taylor 1969). An illustrated review of PAN injury symptoms is available which includes information on relative sensitivity of species (Taylor and MacLean 1970). PAN injury has been found in the Netherlands on lettuce and chrysanthemum, as well as on weed species (Ten Houten 1974). The same injury symptoms could be produced by controlled exposure. The PAN incidence was related to a large refinery and a chemical plant in the vicinity which could provide reactants for PAN synthesis. PAN effects on plants are complex and must include consideration of the chemistry of the peroxyacyl nitrates, including PAN and its homologs, as well as the physiological and biochemical effects of these compounds (Mudd 1975a).

Nitrogen Oxides. Nitrogen oxides (NO_x) are important components of oxidant smog and may also be significant plant toxicants on their own. Detailed information on plant injury due to NO_x is available together with discussion of the chemistry of formation and mode of action of these pollutants (Taylor et al. 1975).

Nitrogen oxides participate in the photochemical reactions resulting in the production of O_3 and PAN. The direct injurious effects of NO_x are generally a result of relatively high concentrations, coming directly from specific industrial sources, for example, nitric acid plants. Nitrogen oxides are the sum of the concentrations of NO and NO_2. Of the many forms of N found in the atmosphere, these are the most significant pollutants emitted by urban and industrial activities. Nitrogen dioxide is much more toxic to plants than NO and reacts with water to form HNO_3 and HNO_2 or NO. Nitric oxide is the major NO_x in combustion emission; its synthesis results from the high temperature exposure of N_2 and O_2 in the combustion air. Significant quantities of NO are formed during the burning of coal, oil, natural gas, or gasoline in power plants or internal combustion engines. After emission, some of the NO is oxidized to NO_2 by reaction with O_2. This conversion is mainly the result of a photochemical reaction of NO_x and hydrocarbons. Thus, NO_2 not only takes part in the photochemical reaction leading to O_3 and PAN production but is also formed from NO in a photochemical reaction. Nitrogen oxides not only mav suppress vegetation growth but also, because they are highly coloured, may contribute to reduced atmospheric visibility in urban areas.

Nitrogen dioxide can cause visible leaf injury at high concentrations and growth reductions at lower concentrations over extended periods (Taylor and Eaton 1966; Spierings 1971). However, NO_2 is seldom found outdoors in sufficiently high concentrations to cause visible injury. It may be important in combination with other pollutants.

Typical of the limited number of NO_x studies, Capron and Mansfield (1975) detected 30 to 40 pphm NO and about 10 pphm NO_2 in glasshouses in which propane and kerosene burners were used for CO_2 enrichment. These concentrations were considered to be below the threshold level for injury from NO_2 alone, but capable of causing injury in the presence of small amounts of SO_2. The possibility that nitrogen oxides may have nutritional value has been investigated. Nitrogen dioxide exposure increases the nitrate content of pea tissue even though leaf necrosis occurs at the concentrations used (Zeevaart 1976). A continuous stirred tank reactor exposure chamber has been used to study the dynamics of NO_2 uptake by corn and soybean (Rogers et al. 1977). Uptake of NO_2 is independent of NO_2 concentration and leaf surface area, but directly dependent upon inverse total diffusion resistance.

Sulphur Dioxide. Sulphur dioxide originates from the combustion of fossil fuels (Fig. 1) and the smelting of S-containing ores. It is incorporated as a nutrient within plant tissue up to a particular concentration, which depends in part on the S-nutrition status, and above which it has toxic effects. Characteristic plant injury symptoms are produced and an illustrated review of plant injury is available which includes relative species susceptibility ratings, injury symptoms, and symptoms of other disorders resembling SO_2 injury (Barrett and Benedict 1970). A thorough colour photographic presentation on SO_2 pollution, SO_2 injury types, species tolerance, and environmental factor effects is also available (Van Haut and Stratmann 1970). The effects of SO_2 on plants and experimental methods for studying them have also been reviewed (Bovay 1969; Daines 1969b). The relationship of SO_2 exposure concentration to plant responses has been summarized (Knabe 1976). Low doses are harmless or even beneficial while increasing concentration causes increasingly severe injury. Knabe (1976) also contrasted the effects of SO_2 and acid precipitation, as well as providing a review of the biochemical effects of SO_2. The biochemical and cytological effects of SO_2 have also been reviewed by Malhotra and Hocking (1976).

The response of plants to SO_2 was characterized in an early series of experiments (Zimmerman and Crocker 1934). Middle-aged leaves are first attacked, interveinal tissues are most sensitive, injury increases with duration of exposure and concentration, wilted plants are more tolerant than turgid plants due, at least in part, to stomatal condition, and plants are more resistant to SO_2 at night. The SO_2 sensitivity of 49 species has been determined using controlled

Fig. 1. Coal-fired electricity generation installation in the Midlands of
England (top). Coal-fired generation station located on a lake-
shore in Ontario, Canada (bottom) (Author's photos).

exposure equipment (Zimmerman and Hitchcock 1956). Wide differences in sensitivity are found which do not relate to stomatal number. A number of additional SO_2-sensitive species have been identified (Van Haut and Stratmann 1970).

Toxicity to plants due to SO_2 is affected by numerous factors including temperature, relative humidity, soil moisture, light intensity, nutrient supply, and age of plant tissue. The physiological and biochemical effects of SO_2 together with its chemical properties have been reviewed (Mudd 1975b). Harmful SO_2 emissions may diffuse and thus be diluted, or may be chemically or physically changed. Sulphur dioxide may oxidize in the ambient atmosphere in the presence of sufficient water to form H_2SO_4 mist which may have a deleterious effect on plants and soils. Acid rain and mist are thus manifestations of SO_2 pollution.

Fluoride. Fluoride pollution results from high temperature treatment of F-containing raw materials, in such processes as Fe and Al refining, superphosphate fertilizer manufacturing, glass manufacturing, and brick making (Fig. 2). For example, chlorotic patterns on citrus leaves were attributed to injury from the F evolved in the process of manufacturing triple-superphosphate (Wander and

Fig. 2.   A brick-making facility from which F may be emitted as a result of high
temperature treatment of F-containing raw materials (Author's photo).

McBride 1956). The rock phosphate raw material contained 2 to 4% F which was released during the acid treatments involved in fertilizer manufacture. The F evolved by various industries becomes phytotoxic at some critical concentration but the F travels a relatively short distance from the source compared to either O_3 and its precursors or SO_2. Severe injury can occur to sensitive species up to a few km from the source. Fluoride accumulates in plants so injury may occur from the accumulative effects of a very low atmospheric F concentration. Fluoride moves in the transpiration stream and may accumulate to quite high concentrations near the tips and margins of the leaves (Treshow and Pack 1970).

Plant responses to F have been thoroughly described, an extensive list of species tolerance ratings has been prepared, many interactions of F with plant factors have been summarized, and a bibliography of 300 references provided (Weinstein 1977). Other reviews include that of Bovay (1969) who included a description of experimental methods. The nature of F injury to plants has also been illustrated and summarized by Treshow and Pack (1970). Species differences were noted as well as environmental effects and the existence of disorders with symptoms similar to F-toxicity. The role of F as an air pollutant affecting plants has been reviewed by Treshow (1971) who provided extensive information on sources, uptake patterns, injury symptoms, environmental interactions, host-parasite relations, and ecological effects. Fluoride influences on a wide range of plant physiological and biochemical processes have been reviewed by Chang (1975).

There have been many studies of species sensitivity to F. A wide range of susceptibility is usually found with the F sensitivity rating not well correlated with SO_2 sensitivity. Fluoride absorption rates differ among species and plant parts but stomatal number is not related to sensitivity. All plants accumulate F but a few like dogwood, deutzia, gifbaar and camellia accumulate abnormally large amounts (Zimmerman and Hitchcock 1956). The average response to HF is less in darkness than in light and plants are more responsive to constant low concentrations than to twice daily higher concentrations according to controlled environment studies (Adams et al. 1957). Plants within each family, genus and species have similar relationships of F exposure to F accumulation.

Exposure of leafy vegetables to HF does not result in growth retardation unless there is visible injury to the leaves. Fluoride entering the leaf is translocated to the margins but not downward in the plant (Benedict et al. 1964). Studies of the accumulation and distribution of F in several species have shown that aerial F is absorbed on leaf surfaces as well as accumulated internally (Jacobson et al. 1966). Internal F is translocated outward to the surface or upward to the tip. Wide differences in sensitivity among species are considered to be due to differences in patterns of accumulation, translocation and distribution of F. Studies of the accumulation of F in fruit, vegetables and other species in the vicinity of an Al plant have shown that F content of fruits ranges from 0.5 to 5.0 mg/kg and of vegetables from 0.5 to 100 mg/kg with lower concentrations in the roots (Rippel and Janovicová 1969).

There have been numerous detailed studies of individual species responses to F. For example, that marginal scorch and leaf spot symptoms on Italian prune trees in the vicinity of Al smelters in Washington State, U.S.A., were due to F toxicity was proven by chemical analysis, and severity of symptoms was proportional to F content of leaves (Miller et al. 1948). When the smelters were closed down the F content of leaves in subsequent growing seasons did not exceed normal levels and injury symptoms did not appear. The relationship of F pollution to leaf necrosis of gladiolus has been proven by both tissue analyses and experimental exposures (Johnson et al. 1950). Other F-induced disorders include soft suture of peach (Benson 1959) and decreased pollen effectiveness in cherry (Facteau et al. 1973). One report includes description of F responses of 49 species (Zimmerman and Hitchcock 1956) while another includes ratings of F sensitivity of 34 species in controlled environment studies (Adams et al. 1957).

Fluoride-induced injury to tomato and corn plants has been characterized (Leone et al. 1956). Short-term high concentration exposures are more injurious and result in more F accumulation and retention than the same amount of HF spread over a longer time. Fluoride accumulation increases with HF exposure concentration and mature leaves accumulate more F than younger leaves. Open-top field chambers have been used to determine HF effects on yield of bean and tomato plants (MacLean et al. 1977). In beans exposure to about 0.6 mg F/m^3 for 43 days reduces pod weight by almost 25%, compared to filtered air, without inducing foliar injury. Tomato is not affected by 99 days exposure to the same concentration of HF.

Penetration of F through pear leaf cuticles has been described (Chamel and Garrec 1977). Penetration of F through intact cuticles is very slow and penetration under natural conditions would be facilitated by breaks or punctures in the cuticle. Experiments have been conducted which show that plants can convert some F to a volatile form which may be lost from the plant and be absorbed in solution (Peters and Shorthouse 1967). Appreciable amounts of F can be washed from plant foliage by rain, reducing leaf injury (Treshow and Pack 1970). Studies of the localization of F accumulation in plant cells have shown that this element is present, in decreasing order, in chloroplasts, cell wall, water soluble protein, and mitochondria (Chang and Thompson 1965). Almost 60% of the F is in the supernatant fraction.

There are many problems involved in interpreting F effects on plants
including the complex relationship between atmospheric F concentration and
visible injury (Hill 1969; McCune 1969). Also injury is associated with
accumulation over a long-term exposure, and species and cultivars differ greatly
in F susceptibility. Gaseous forms of F are the most important pollutants, and
air monitoring techniques must discriminate between particulate and gaseous
forms. Fluoride-induced effects differ, species and stages of development differ
in susceptibility, and environmental factors affect response. Injury assessment
must consider both foliar markings and reduction in growth.

Ethylene. Ethylene, the unsaturated hydrocarbon C_2H_4 is, at very low concentrations, a natural growth regulator in plants. Ethylene concentrations in
plants may be manipulated by the application of ethephon, a compound which
releases ethylene in the tissue, accelerating ripening and hastening abscission
and senescence. Maturing fruit may release so much ethylene while ripening in
storage, that sensitive plants growing nearby or stored in the same facility
are injured.

Potentially injurious ethylene air pollution is present in urban centres
on a continual basis and constitutes a stress on growing and developing plants
(Abeles and Heggestad 1973). The primary sources of this pollutant are motor
vehicle exhaust and leakage from some chemical factories, such as those of the
plastics industries. Mixtures of gases from such sources often cause typical
ethylene injury symptoms. For example, combinations of ethylene, acetylene,
propylene and nitrogen dioxide cause ethylene injury symptoms on plants (Heck
1964). At one time illuminating gas was a source of troublesome amounts of
ethylene, but this problem disappeared with the advent of electric lighting.

Ethylene pollution has a substantial impact on floriculture. The sources
are the plants themselves especially ripening fruit and injured tissue, fungi,
decaying organic matter, burning organic matter, industrial activities, heaters,
engines, fuels and oils. Various management practices designed to ameliorate
ethylene injury are recommended for glasshouses, refrigerators and transportation facilities (Hasek et al. 1969).

Glasshouse CO_2 enrichment by means of natural gas burners may result in
injury to sensitive species by the ethylene produced by poorly ventilated
burners (Hanan 1973a). Ethylene may reach injurious levels unless the burners
are equipped with a distribution tube extending the length of the glasshouse to
force complete air circulation with fresh air brought to the burner. Specific
provision must be made in greenhouse structures for combustion air entry,
especially when outside temperatures are so low that ice seals leaks in the
glasshouse structure, or when an outer layer of plastic film covers the
glasshouse.

Other Gaseous Pollutants. The concentrations of other gaseous pollutants are normally too low to cause injury symptoms in plants. Potential pollutants such as CO, Cl_2, NH_3, HCl, Hg and H_2S are infrequent problems and any injury is usually confined to a limited region under special circumstances. Colour photographs of injury due to many of these pollutants are available along with species lists of relative sensitivity (Heck et al. 1970). Seeds are relatively insensitive to such air pollutants as Cl_2, HCN, H_2S, NH_3 and SO_2 (Barton 1940).

Carbon monoxide. Even though CO is considered to be a serious pollutant of the urban atmosphere (Hexter and Goldsmith 1971), its effects on horticultural plants have not been reported. Plant leaves do absorb considerable amounts of CO (Bidwell and Fraser 1972) but possible concurrent injury symptoms have not been described. It is apparent that injury due to CO is not anticipated at the average concentrations found in urban atmospheres.

Chlorine. Accidental spills of Cl_2 may be injurious in localized areas. Plant sensitivity to Cl_2 has been characterized using controlled exposures with observations corroborated by injury symptoms on plants exposed outdoors to an accidental release of Cl_2 which injured many species (Brennan et al. 1965, 1969). Another case of vegetation injury due to accidental release of Cl_2 gas has been used to characterize the symptomatology and rate the sensitivity of a wide variety of ornamental trees and shrubs (Rhoads and Brennan 1976). The episode served to pinpoint Cl_2 effects which, if the cause had not been known, would have been difficult to diagnose because of variations in symptoms among species and similarity of symptoms to injury by other agents.

Ammonia. Plants may be injured by NH_3 in the vicinity of an accidental spill or near large numbers of farm animals. Ammonia injury may occur to stored produce if the gas is released accidentally from the refrigeration system. Injury symptoms on apple and peach fruit have been verified by controlled exposure to NH_3 (Brennan et al. 1962). Analysis for total N showed that this parameter could not be used as an indicator of accidental exposure.

Hydrogen chloride. Some localized plant injury may result from the burning of polyvinyl chloride plastics releasing HCl.

Hydrogen sulfide. Young plant tissue is generally more sensitive to H_2S than older; injury symptoms are necrosis of young shoots and leaves, and basal and marginal necrosis of the next oldest leaves; injury increases rapidly with increases in temperature; wilted plants are less sensitive than unwilted (McCallan et al. 1936). Apple, cherry, peach and strawberry are relatively resistant to H_2S. Pepper, gladiolus and sunflower have slight to moderate injury, and cucumber, tomato and radish are relatively sensitive. Hydrogen sulfide effects on photosynthesis, respiration and water relations of spinach have been studied in specially designed exposure facilities (Oliva and Steubing 1976).

Unidentified gases. Occasionally apparent air pollution injury not readily attributable to any known gaseous pollutant is reported. For example, outdoor injury symptoms differing from those due to O_3 or SO_2 were described (Rich and Tomlinson 1968). The particular symptoms on petunia were referred to as "aldehyde" injury. Studies involving unfiltered internal combustion engine exhaust gases have sometimes not included a quantitative assessment of gas composition. For example, injury to redbud as a result of close proximity to automobile exhaust fumes has been described (Parris 1968). The dieback symptoms are associated with cool high humidity conditions in spring.

Air Pollution Intrusion into Buildings. The concentrations of pollutants entering buildings will be of importance in the culture of indoor decorative plants. Studies of the relationship of outdoor to indoor levels of pollutants indicate that, in general, indoor concentrations depend upon how much outside air is brought in (Thompson et al. 1973). With rapid intake, levels may be 2/3 those outside; with little intake the amounts inside may be near zero. Nitrogen oxides and CO persist longer indoors than O_3 or PAN. Particulate matter levels are also directly related to ventilation rate. All air pollutants are reduced most indoors compared to outdoors in buildings with refrigerated air conditioning systems, followed by evaporative coolers, with no air conditioning ventilation the least effective.

Particulates. Particulates may have a role as plant-injuring air pollutants. Large amounts of particulates are emitted according to a study of particulate air pollution emissions in the U.S.A. (Vandegrift et al. 1971). The leading sources are power generation plants, stone crushing, agriculture and related operations, cement, iron and steel manufacture, and forest product processing. The effects of particulates on plants have been summarized (Lerman and Darley 1975). Cement-kiln dusts are occasionally deleterious to plants, especially if deposited on wet leaves so that crusting occurs. The probable causes of injury are considered to be screening of light, plugging of stomata, and chemical reactions on the leaf surface. The composition of cement-kiln dust is highly variable. Particulate fluorides are less harmful than gaseous forms. Soot, magnesium oxide, iron oxide, and lead are particulates ranging widely in phytotoxicity.

A detailed study of injury to vegetation from cement-kiln dust has shown that injury is a result of a combination of a relatively thick crust deposit and the toxicity of alkaline solutions formed when dusts are deposited in the presence of water (Darley 1966). Finer particles of some cement-kiln dusts apparently interfere with gaseous exchange and may cause leaf injury. Calcium content alone is not always a good indicator of toxicity. Chemical composition, particle size, and deposition rate all affect phytotoxicity. Soot particles from thermal electric generation units caused necrotic spotting of beans, cucumbers and African violets growing in greenhouses (Miller and Rich 1967). The soot particles had a pH of 2.0 and 8.5% titrable acid, probably from S in the fuel. Plants outside the greenhouse were not injured, possibly because of rain washing the foliage.

Trace element particulates such as Pb, Zn and Cd may become airborne and contaminate vegetation through the leaves and other above-ground surfaces. Salt sprays near oceans and from salt used for ice control on roadways may also constitute phytotoxic air pollutants. These two forms of pollution are considered in detail elsewhere in this volume.

Acid Precipitation. Acid precipitation may occur when gaseous pollutants such as SO_2 and NO_x are converted chemically in the atmosphere to strong acids. Effects on vegetation are a result of wet deposition of acidic gases and particles as part of rain and snow (Knabe 1976). The effects on vegetation are considered to be mainly indirect and attributable to chemical or biological changes in the soil. The formation and distribution of acid rains on a world-wide basis has been reviewed with the problem areas identified as northern Europe and north-eastern U.S.A. (Likens et al. 1972).

Acid mist effects on yellow birch seedlings have been described (Wood and Bormann 1974). Foliar tissue injury is observed at pH 3.0 or below, with significant growth reduction at pH 2.3. Apparent photosynthetic rates increase dramatically when simulated acid rain is applied to bean leaves (Ferenbaugh 1976), while respiration increases only slightly in response to pH below 3 (Table 2). Leaves of affected plants are characterized by smaller cells, less intercellular space, and smaller starch granules.

Table 2. Chlorophyll content, respiration rate, and photosynthesis rate for control and acid rain-treated bean leaves.

	Chlorophyll mg/g fresh wt	Respiration $\mu l\ O_2$/g fresh wt/h	Photosynthesis $\mu l\ O_2$/g fresh wt/h
Control	1.96	445	230
pH 3.5 acid rain	1.87	438	467
pH 3.0	2.02	495	544
pH 2.5	1.86	598	676
pH 2.0	1.24	529	860

Data from Ferenbaugh (1976)

White pine seedling productivity increases with decrease in pH of simulated acid rain from 5.6 to 2.3 in spite of near depletion of available K, Mg and Ca ions in the soil at pH 2.3 (Wood and Bormann 1977). Additions of HNO_3 to the simulated rain are associated with increased foliar N concentrations.

Air Pollutant Mixtures. Plants are usually exposed to mixtures of pollutants under outdoor conditions because the natural environment generally contains a mixture of potential plant-injuring gases. The effects of air pollutants in mixtures on horticultural plants is not well understood and plant responses to combined pollutants may differ from separate exposure effects. The greatest concern relates to the observation that the combined effects of two or more gases may lower the threshold levels of the individual contaminants required to cause injury when each component alone at the same concentration does not induce visible symptoms (Fig. 3). There is thus frequent definite enhancement of

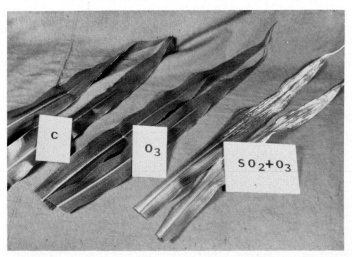

Fig. 3.   Leaves from corn plants exposed to O_3 and O_3 + SO_2 compared to control (C). SO_2-treated plants had leaves similar to the control (Author's photo).

response to multiple pollutants compared to single gases, but little is known of the mode of action of mixtures. Our present knowledge of mixture effects is very fragmentary and much remains to be done with ranges of concentration, ratios of pollutants, mixtures of more than two pollutants, age and physiological condition of plants, and environmental conditions before, during and after exposure.

A background on the concepts involved in responses of plants to pollutant combinations has been developed along with a compilation of air pollutant mixture research findings on various crops including some of horticultural significance (Reinert et al. 1975). The studies conducted to date have indicated that pollutant mixtures can have important effects on vegetables, fruit and ornamental crops. For example, exposure of bean, tomato and radish to NO_2 and SO_2 mixtures results in foliage injury at much lower concentration for mixtures than for either pollutant alone (Tingey et al. 1971b). The ratio of combined pollutants appears to be very important in determining injury (Table 3).

Table 3. Injury to the upper leaf surface of plants exposed to mixtures of NO_2 and SO_2 for 4 hr.[y]

NO_2 pphm	SO_2 pphm	Leaf injury[z] (%)		
		Pinto bean	Radish	Tomato
5	5	2	1	0
5	10	0	0	0
5	20	1	0	0
5	25	1	0	1
10	5	0	1	0
10	10	11	27	1
15	10	24	24	17
15	25	4	4	0
20	20	16	6	0
25	5	0	13	0

[y] Threshold injury concentrations for NO_2 and SO_2 in 4 hr exposures were 200 and 50 pphm, respectively.

[z] Average % leaf injury to the three most severely injured leaves/plant except for bean, which was based on injury to the two primary leaves.

Data from Tingey et al. (1971b)

Ambient pollutant monitoring records illustrate that the concentrations studied experimentally are realistic in relation to ambient concentrations (Table 4).

Exposure of several vegetable species to mixtures of O_3 and SO_2 at concentrations similar to those found in urban areas indicates that more than additive injury occurs in radish, while in cabbage and tomato the foliar injury is additive or less than additive (Table 5). Plants exposed to mixtures of O_3 and SO_2 usually have injury symptoms resembling those due to O_3 alone while some plants have symptoms resembling those caused by SO_2.

Evaluation of effects of SO_2 and/or O_3 on eastern white pine using 5 pphm

Table 4. Atmospheric concentrations of NO_2 and SO_2 in selected metropolitan areas in the U.S.A. exceeded 1, 10 and 30% of the time.

City	NO_2 1962–68			SO_2 1962–67		
	1%	10%	30%	1%	10%	30%
Chicago	12[z]	7	5	65	33	16
Cincinnati	9	5	4	17	7	3
Denver	10	6	4	6	3	2
Los Angeles	24	11	6	8	4	2
Philadelphia	11	7	4	45	21	9
St. Louis	8	5	3	25	11	5
San Francisco	18	8	6	7	3	1
Washington, D.C.	10	6	4	21	10	6

[z] Average hourly concentrations (pphm).

Data from Tingey et al. (1971b)

Table 5. Summary of effects of O_3/SO_2 interactions on foliar injury.

Species	O_3/SO_2 concentration (pphm)			
	5/50	10/10	10/25	10/50
Broccoli	+	+	0	0
Cabbage	0	0	0	+
Radish	0	+	+	+
Tomato	0	–	0	0

0: injury from mix equal to additive injury from single gases.

+: injury from mix greater than additive injury from single gases.

–: injury from mix less than additive injury from single gases.

Data from Tingey et al. (1973d)

of each gas showed that there is less injury to young needles from the SO_2-O_3 mixture than from SO_2 alone and no injury from O_3 alone (Costonis 1973). Greatest injury results from a separate exposure to O_3 and SO_2 followed 24 hr later by exposure to the mixture of gases.

Other studies of gas mixture effects include O_3 and SO_2 on apple trees (Kender and Spierings 1975), and beans (Hofstra and Ormrod 1977), SO_2 and HF on citrus seedlings (Matsushima and Brewer 1972), SO_2 and NO_2 on vegetable crops (Bennett et al. 1975), SO_2 and HF on bean and sweet corn (Mandl et al. 1975), O_3 and PAN on several species (Posthumus 1977), and O_3, SO_2 and NO_2 on beans (Ormrod and Hofstra, unpublished, Fig. 4).

Fig. 4.   Bean leaf injury resulting from exposure to mixed gases (Author's
          photo).

Many studies of air pollutant mixture effects have included more than one
species in the same experiment. This allows direct comparison of species responses and greater assurance of precision in such comparisons. For example, in
a study of the effects of combinations of SO_2 and NO_2 on several vegetable crops
it was found that the combination may enhance the phytotoxicity of these pollutants but relatively high doses are required to cause visible injury (Bennett
et al. 1975). The minimum concentration required for visible injury to radish
leaves is 75 pphm SO_2 for one hr or 50 pphm SO_2 combined with 50 pphm NO_2.
Swiss chard and peas require 75 to 100 pphm SO_2 alone or in combination with
the same concentration of NO_2 to exhibit leaf injury. In another comparative
study, sweet corn and bean were exposed to SO_2 and HF (Mandl et al. 1975).
Exposure to 15 or 30 pphm SO_2 injures sweet corn leaves but not bean leaves.
Combination with HF does not alter the sweet corn response. At six to eight
pphm SO_2 the foliar injury of bean is accentuated by combination with HF.

1.6   Interactions of Air Pollutants with Other Pollutants in the Environment

Trace Elements and Air Pollutants. The presence of trace elements may
affect plant response to air pollutants. Trace elements and O_3 interact with
enhancement of injury when both pollutants are present. Application of Cd or
Zn solutions to leaves or rooting media of cress or lettuce increases O_3-induced

phytotoxicity (Czuba and Ormrod). The concentrations of heavy metal solutions used are below those resulting in metal-induced growth retardation or leaf necrosis. Such an interaction of the metals Cd and Ni with O_3 can be demonstrated for peas (Ormrod 1977). A range of metal concentration in the tissue, not itself suppressing growth, enhances O_3-induced phytotoxicity. If trace element-induced growth-retarding concentrations are used on peas, the plants are much more insensitive to O_3 than control plants. This response also occurs in 'Pinto' beans. Ozone sensitivity increases as available soil Zn increases (McIlveen et al. 1975).

The combined effects of heavy metals and SO_2 on several species have also been studied (Krause and Kaiser 1977). Sulphur dioxide does not influence metal uptake or translocation, or heavy metal-induced yield reduction of plants dusted with a mixture of CdO, PbO_2, CuO and MnO_2, but foliar injury induced by heavy metals is increased by SO_2 in all plants.

Pesticides and Air Pollutants. Air pollutants may interact with selective herbicides to give unexpected crop injury. Ozone can interact with herbicides to modify over all plant response to the herbicide or its metabolism. There are greater than additive reactions of O_3 and some herbicides in suppressing plant growth (Carney et al. 1973) but other herbicides do not interact with O_3. The time course of diphenamid metabolism in tomato is altered by exposure to O_3 (Hodgson et al. 1973, 1974). This O_3 effect was utilized to elucidate a scheme for diphenamid metabolism in the plant. In pepper, O_3 does not affect diphenamid absorption or translocation but does enhance accumulation of some water-soluble metabolites in leaves while decreasing them in roots (Hodgson and Hoffer 1977a). Ozone exposure stimulates the production of two types of glycoside conjugates, both metabolites of diphenamid (Hodgson and Hoffer 1977b). There may be an antagonistic action between O_3 and 2,4-D when both are air pollutants (Sherwood and Rolph 1970). Tomato, zinnia and elm plants survive otherwise injurious dosages of 2,4-D and O_3 if the two are applied simultaneously or almost so.

Ozone sensitivity of 'Pinto' bean plants is increased by addition of the contact nematicides penamiphos, fensulfothion, aldicarb and oxamyl (Miller et al. 1976).

Salinity and Ozone. High rooting medium salinity results in decreased garden beet yield and also decreased O_3 injury (Ogata and Maas 1973). Similarly, O_3 injury to bean decreases with increasing salinity (Hoffman et al. 1973).

Plant Parasites and Air Pollutants. The interactions between air pollutants and plant parasites have been thoroughly reviewed (Heagle 1973) with separate consideration of SO_2, O_3 and F effects on fungus diseases; and more general reviews of interactions of pollutants with bacteria, pollutants with viruses, and pollutants with insects. There are many aspects to the inter-actions of air pollutants and plant diseases (Treshow 1975). Both the impact of pollutants on pathogens and the impact of diseases on pollutant response must be considered, as well as possible modes of action. Ozone-injured plants are more susceptible to facultative parasitic fungi but less so to obligate parasites (Manning 1975). Sulphur dioxide may retard microorganism activity directly while F can have a similar effect depending on concentration. There are numerous examples of such interactions involving horticultural crops.

Oxidants. Rust-infected leaves of bean and sunflower are less injured by oxidant smog than healthy leaves (Yarwood and Middleton 1954). Oxidant-protection is associated with some substance that diffuses beyond the limits of the rust mycelium and is not associated with stomatal closure.

Infection of potato leaves by Botrytis cinerea is more rapid and severe in O_3-injured leaves than in uninjured leaves (Manning et al. 1969). Infection often begins in the O_3-injured areas of the leaf. Similar effects are observed in powdery mildew infection of pumpkin (Fig. 5). In geranium previous exposure to O_3 does not appreciably influence the susceptibility of leaves to infection by B. cinerea, unless there is visible O_3 injury (Manning et al. 1970). Ozone-injured, necrotic tissues on older attached and detached geranium leaves serve as infection sites for B. cinerea. When flower bracts of poinsettia plants are

Fig. 5. Apparent O_3 injury to pumpkin leaves in the field is followed by severe infection with powdery mildew (Author's photo).

inoculated with B. cinerea and exposed to O_3 there is no effect on infectivity (Manning et al. 1972a). Apparent increases in Botrytis injury on onions with increased O_3 injury can be demonstrated by applying fungicide-antiozonant combinations (Wukasch and Hofstra 1977a) or by using carbon filter-equipped open-top field chambers (Wukasch and Hofstra 1977b).

Infectivity of Fusarium oxysporum f. sp. conglutinians on cabbage is not affected by exposure of plants to 10 - 12 pphm O_3, eight hr per day, for 10 weeks (Manning et al. 1971a). In contrast, 'Pinto' bean plants exposed to chronic levels of O_3 have more root surface fungi and fewer Rhizobium nodules than control plants (Manning et al. 1971b).

Interaction of virus infection with O_3 response of 'Pinto' bean can be demonstrated (Davis and Smith 1974, 1975). Plants inoculated with bean common mosaic virus five days previous to O_3 exposure have less O_3 injury than non-infected control plants. Infecting 'Pinto' bean plants with alfalfa mosaic, tobacco ringspot, tomato ringspot, or tobacco mosaic viruses prior to O_3 treatment reduces O_3 injury with O_3 sensitivity directly related to concentration of virus inoculum (Davis and Smith 1976). Inoculation of one leaf on a plant is sufficient to reduce injury in the opposite non-injured leaf. However, inoculation of tomato with virus seven days before O_3 exposure enhances O_3 injury (Fig. 6).

There is a superficial similarity between the white flecking caused by mesophyll-feeding leafhoppers and that caused by O_3. Examination of both effects on several landscape trees reveals differences in location, configuration, colour, and size of external symptoms, and different characteristics of parenchyma tissue injury (Hibben 1969b).

Sulphur dioxide. Sulphur dioxide at a concentration of 100 $\mu g/m^3$ for 2 days markedly reduces the infectivity of black spot (Diplocapon rosae) on rose foliage (Saunders 1966). In areas where SO_2 pollution levels exceeds 100 $\mu g/m^3$, the disease is checked or eliminated. Sulphur dioxide decreases the incidence and severity of bean rust (Weinstein et al. 1975).

The spider mite, Partetranychus pilosum is susceptible to SO_2 exposure in a polluted area but Aphis pomella is highly resistant (Przybylski 1967). Thus the population of A. pomella in an orchard may increase while that of P. pilosum decreases.

Fluoride. Fluoride affects tobacco mosaic virus lesion development on 'Pinto' beans (Treshow et al. 1967). The number of lesions increases with foliar F up to 500 ppm, above which the number decreases. There are fewer lesions on older leaves.

Fig. 6. Ozone injury on tomato plants resulting from the interaction of virus
infection with O_3 exposure. Plants were inoculated with CMV, TMV, or
CMV + TMV 7 days before O_3 exposures (Kemp and Ormrod unpublished).

1.7 Absorption of Air Pollutants

The nature of the interaction of plant canopies with major gaseous pollutants in the immediate vicinity of the plants is complex (Bennett and Hill
1975). Factors that must be considered are gaseous pollutant interchanges
between plant canopies and adjacent boundary layers, as well as potential plant
sinks and pollutant fate within the vegetation. The rates of uptake of a number of pollutants increase as their solubility in water increases (Hill 1971).
The exceptions are O_3 and PAN which, because of their high degree of chemical
reactivity, have higher uptake rates than would be expected on the basis of
their solubilities. Stomata have important roles in determining the pollutant
absorption and response of plants (Mansfield 1973). Stomata help to determine
susceptibility through their role as the principal port of entry with apertures
affected by environmental conditions, as for example, in the closing of stomata
at low humidity.

Stomata may also respond directly to air pollutants, for example, the
stimulation of opening by SO_2 at high humidity, and thus affect the gaseous

exchange rates of the plant. The increased CO_2 concentrations often associated with air pollution episodes also may have a role (Mansfield and Majernik 1970). There is considerable information on the effects of air pollutants on gaseous exchange (Unsworth et al. 1976). The design and analysis of experiments on gas exchange using SO_2 is an example. Resistance analogues are used to separate gaseous diffusion effects from internal resistance effects.

Oxidants. Demonstrations of stomatal control of O_3 uptake show that factors that control transpiration also control O_3 uptake (Dugger et al. 1962; Rich et al. 1970). Ozone affects stomatal resistance of 'Pinto' beans under different soil water and atmospheric humidity conditions (Rich and Turner 1972). Stomatal opening responses depend on both soil water stress and atmospheric moisture. Darkness, soil-plant water stress and PMA close stomata which, in turn, protect plants against O_3 (Adedipe et al. 1973).

The inheritance and mechanism of action of O_3 resistance in onion have been clearly defined (Engle and Gabelman 1966). A dominant genetic system regulates stomate sensitivity to O_3. Guard cell membranes of O_3-resistant plants lose permeability and leak in the presence of O_3, closing the stomata and preventing further entry of O_3 into the leaves.

Lenticels are the gaseous exchange parts for many fruit and stems. There has been little study of their response to pollutants and role in pollutant uptake. They are affected by O_3 in apple fruit with the formation of pitted areas around lenticels after O_3 exposure (Miller and Rich 1968). Ammonia also causes lenticular spotting (Brennan et al. 1962). The lenticular tissue may be the most exposed to pollutants of any fruit or stem surface tissue.

Sulphur Dioxide. There have been many studies of SO_2 effects on stomatal responses (Ziegler 1973b; Verkroost 1974). Details of the SO_2 stimulation of stomate opening in broad bean reveal the response pattern (Majernik and Mansfield 1971). Sulphur dioxide in the range 0.25 to 9.0 ppm causes stomata to be much wider open than on control plants. Night closure is not much affected. The wider opening could lead to excessive transpiration and water stress, and easier entry to the leaf of SO_2 and other pollutants. Stomatal closing occurs in broad bean in the presence of elevated CO_2 (Majernik and Mansfield 1972) even though SO_2 is present in sufficient quantities to induce stomatal opening. Stimulation of stomatal opening by SO_2 in broad bean is considerably greater in older leaves compared with younger leaves (Biscoe et al. 1973) and SO_2 exposure increases the transpiration rate between 23 and 32%, depending on growing conditions. High humidity or removal of atmospheric CO_2 increases stomatal opening of geranium leaves resulting in increased SO_2-induced necrosis (Bonte and Longuet 1975). The SO_2, however, causes temporary stomatal closure in this species.

1.8 Pollutant Removal

The ability of some plants to remove significant quantities of pollutants from the air without sustaining serious foliar injury or growth retardation has horticultural significance (Waggoner 1972). Experiments on pollutant uptake by shade trees show that conifers are more effective than deciduous hardwood trees in removing particulate pollutants from the atmosphere (Dochinger 1972).

Oxidants. Removal rates of O_3 indicate that white oak and white birch are effective O_3 scrubbers while red maple and white ash are relatively inefficient in removing O_3 (Table 6).

Table 6. Rates of O_3 uptake by 9 shade tree species from an atmosphere containing 20 pphm O_3.

Species	No. of seedlings sampled	Uptake rate	
		mg O_3dm^{-2}hr^{-1}	mg O_3g^{-1}hr^{-1}
White oak	16	0.635a[z]	1.318b
White birch	42	0.536ab	2.347a
Coliseum maple	28	0.502b	0.991c
Sugar maple	16	0.371c	0.863c
Ohio buckeye	16	0.362c	0.927c
Redvein maple	28	0.285cd	0.911c
Sweetgum	44	0.278cd	0.854c
Red maple	88	0.272cd	0.555d
White ash	20	0.239d	0.562d

[z] Mean separation within columns by Duncan's multiple range test, 5% level.
Data from Townsend (1974)

Herbaceous ornamental plants absorb more O_2 than woody species (Table 7), and actively growing woody tissues are much more efficient absorbers of pollutants than non-growing tissue (Table 8).

Atmospheric O_3 is removed rapidly by soil (Turner et al. 1973). Compaction and increasing soil moisture decrease O_3 uptake by the soil.

Sulphur Dioxide. Removal rates of atmospheric SO_2 by foliage have been calculated (Martin and Barber 1971). Other studies of foliar SO_2 absorption by woody plant leaves have been reported (Roberts 1974; Roberts and Krause 1976).

1.9 Environment Interactions

Air pollutant injury to plants is modified by several environmental factors, prominent among which are temperature, light and humidity (Heck 1968; Heck et al. 1965; Rich 1964). Others have added edaphic and nutritional

Table 7. Ozone deposition rates for various species tested under laboratory conditions.

Species	$Vg^y \pm SE$ (cm min^{-1})	Number of tests
Herbaceous		
Petunia	13.7 ± 0.5	11
Osteospermum	12.8 ± 1.0	12
Chrysanthemum	9.8 ± 0.6	14
Woody		
Camellia NG[z]	4.4 ± 0.2	3
Bougainvillea	3.0 ± 0.2	18
Ginkgo	2.0 ± 0.3	4
Quercus	1.9 ± 0.2	17
Camellia OG[z]	1.5 ± 0.1	7
Control - rough paper	2.8 ± 0.8	5

[y] $Vg = \dfrac{O_3 \text{ sorbed/min/cm}^2 \text{ of leaf surface}}{O_3 \text{ concentration in } \mu l/cm^3}$

[z] NG and OG refer to plants with young vigorously growing leaves or with entirely mature growth, respectively.

Data from Thorne and Hanson (1972)

Table 8. Effect of leaf age on rates of O_3 deposition and transpiration in tomato.

Experiment No.	Leaf position[y]	$\overline{Vg}_o{}^z$ (cm min^{-1})	Transpiration ($\mu l/cm^2/min$)
1	3	17.8	0.95
	5	17.6	0.64
	7	8.7	0.57
2	3	9.1	0.93
	5	6.7	0.77
	7	3.9	0.53
3	3	10.6	0.89
	6	6.8	0.49
	7	8.4	0.55

[y] Position from the top of the plant; 7 is oldest.

[z] Average deposition rate.

Data from Thorne and Hanson (1972)

factors to the list of environmental parameters (Taylor 1974). The effects of these environmental variables should be known to permit the development of models which can provide an understanding of the importance of environmental factors in relation to genetic and other plant factors. While much information is available, it is fragmentary, and a large comprehensive research programme would be required to evaluate environmental interactions fully.

Environmental factors may even change the classification of plants as sensitive or insensitive. The relationship of pollutant response to environment is often complex and may differ among species and cultivars. Temperature variations during growth prior to exposure have been found to affect plant sensitivity, but there are few universal response patterns.

1.10 Light

Tomato plants are much more sensitive to O_3 in light than in darkness (Fig. 7). This response likely reflects closure of the stomata in the dark

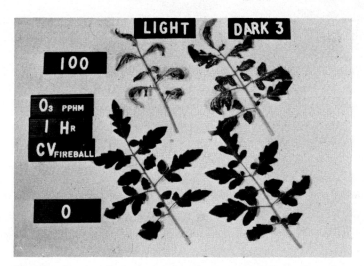

Fig. 7. Tomato plants exposed to O_3 in the dark (upper right) are much less injured than those exposed in the light (upper left) (Adedipe et al. 1973).

and the response would be similar for all species with stomatal closure in the dark. Plants are also more sensitive to SO_2 in the day than in the night (Zimmerman and Crocker 1934). Average response to HF in darkness is 91% of that in light according to a test of 34 plant species in controlled exposure facilities (Adams et al. 1957).

There are effects of exposure time within the daily light period. Injury by synthetic smog on tomatoes is found only at mid-day (Koritz and Went 1953) or in the afternoon (Reinert et al. 1972a) (Fig. 8). Increased water supply

Fig. 8. Tomato plants exposed to O_3 at 3:00 p.m. (3A) near the middle of the
photoperiod were more injured than those exposed at 12:00 noon (12N)
or 6:00 p.m. (6E), and those exposed at 9:00 a.m. (9M) or 9:00 p.m.
(9E) were resistant even though lighted (Adedipe and Ormrod,
unpublished).

also increases oxidant smog susceptibility. These patterns likely reflect
stomatal opening responses to light and water status respectively.

Some effects of post-exposure light conditions have been identified. For
example, a short exposure of 'Pinto' bean plants to PAN requires a two to four
hr 'post light' period for injury development (Taylor et al. 1961).

Light intensity effects have also been found. Plants exposed at medium or
high light intensity are generally more sensitive to O_3 than plants at low
light intensity (Juhren et al. 1957). 'Pinto' bean O_3-induced leaf injury is
significantly increased by higher light intensity and higher relative humidity
both before and during exposure (Dunning and Heck 1973). At high light intensity, however, there is somewhat less leaf injury at high than low humidity.
Inclusion of temperature along with light intensity and humidity reveals that
exposure light intensity interacts with growth light intensity, and growth
temperature with exposure temperature (Dunning and Heck 1977). Bean plants
grown at higher light intensity are less sensitive to O_3 while those grown at
highest temperature or humidity are most sensitive to O_3. Higher exposure
temperature, light and humidity generally result in greatest injury.

Susceptibility of leaves to NO_2 injury increases in low light conditions or darkness (Taylor 1973). For example, about one-half the amount of NO_2 is required in the atmosphere to produce comparable injury on a dark cloudy day compared with a bright sunny day. This principle also applies to light-dark comparisons. There is more leaf injury in the dark than in the light when peas are exposed to NO_2 (Zeevaart 1976).

1.11 Temperature

Generally, temperatures that promote good growth also predispose plants to greater pollutant injury (Heck et al. 1965; Juhren et al. 1957; Taylor et al. 1960). Reduced sensitivity of spinach and lettuce to several pollutants was obtained by reducing temperature (Kendrick et al. 1953). Ethylene sensitivity of carnation flowers is increased by increasing length of cool storage (Mayak and Kofranek 1976).

Temperature effects on sensitivity of annual bluegrass to oxidant smog were investigated as part of a program to standardize the use of this species as an indicator of oxidant smog (Juhren et al. 1957). Plants grown at high temperature are sensitive sooner than those grown at low temperature but the sensitivity diminishes to insensitivity more quickly. Romaine lettuce is less sensitive to oxidants at $13^{\circ}C$ than at $24^{\circ}C$ (Kendrick et al. 1953). Foliar growth of radish is decreased more by O_3 at warmer than cooler temperatures but there is no temperature X O_3 interaction affecting root growth (Tingey et al. 1973a).

Pre- and post-exposure temperatures affect O_3 response of radish (Adedipe and Ormrod 1974). The response pattern differs markedly between cultivars (Fig. 9). In one cultivar low pre-exposure temperature enhances O_3-induced growth suppression, while in the other the same result is obtained by low post-exposure temperature. The temperatures used have no over all effect on plant growth. Greater susceptibility in this species is therefore not related to optimum temperature for growth, in contrast to the response of many other species (Darley and Middleton 1966).

Temperature interacts with S nutrition to affect O_3 response of beans, with protection by high S nutrition at lower temperatures (Adedipe et al. 1972b). Temperature also modifies the O_3 responses of radish plants held at different levels of N and P nutrition (Ormrod et al. 1973). There is greater O_3 suppression of radish dry weight at $20/15^{\circ}C$ (day/night) than at $30/25^{\circ}C$.

1.12 Water Status

Generally, plants grown under water stress conditions are relatively insensitive to atmospheric pollutants, indicating that an increase in soil

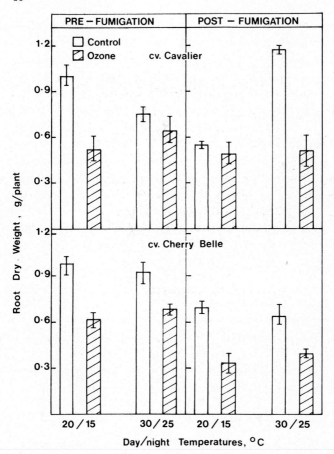

Fig. 9. Ozone effect on root weight of radish plants. Means of 12 replicates ± standard error (Adedipe and Ormrod 1974).

water, which increases plant water status, also increases air pollutant injury. Under field conditions, therefore, it may be expected that loss of irrigated crop value resulting from air pollutant injury could be considerably reduced, if not prevented, by following a less frequent irrigation schedule (but not to the point of adversely affecting crop yield by drought stress), or by temporarily withholding water during air pollution episodes (Middleton 1956). As for light effects, the water effects are probably largely a function of stomatal opening.

Petunia and 'Pinto' beans are protected against irradiated automobile exhaust by drought conditions (Seidman et al. 1965). Excessive irrigation enhances speckle leaf disorder of potato leaves, an oxidant response (Vitosh and Chase 1973).

Near optimum soil water content results in O_3-induced growth reductions in tomatoes (Khatamian et al. 1973). Soil water stress not sufficient to reduce growth due to plant water stress, but resulting in a low relative turgidity of 80% prior to exposure, protects against O_3 (Fig. 10). The

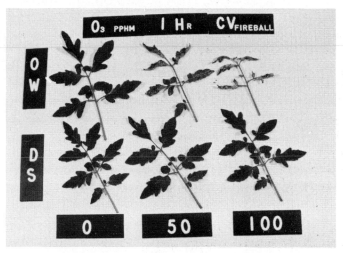

Fig. 10.   Tomato plants watered frequently (OW) were much more sensitive to O_3 than plants watered less frequently (DS) even though the latter treatment did not retard growth (Khatamian et al. 1973).

effects are mainly restricted to soil-plant-water conditions prior to and during O_3 treatment. Ozone effects are therefore those related to tissue water contents per se, independent of growth. Stomatal opening is affected by soil water stress and, in turn, regulates O_3 phytotoxicity (Adedipe et al. 1973).

High atmospheric humidity is usually associated with more severe air pollu-tant injury than low humidity (Heck et al. 1965). This response is also probably largely related to stomate opening response. High relative humidity associated with fan and pad cooling of glasshouses predisposes plants to more severe O_3 injury (Otto and Daines 1969). Sensitivity of begonias to O_3 or SO_2 increases with increasing humidity (Leone and Brennan 1969).

The nutrient solution salt concentration has an effect on oxidant smog injury to sunflower (Oertli 1959). High osmotic potential-induced water stress results in less oxidant injury, a response probably related to the closure of stomata at high plant water stress. Reduction of soil O_2 diffusion rate reduces the O_3 susceptibility of tomato plants growing in ventilated root chambers (Stolzy et al. 1961). This response pattern may be related to reduced water uptake with low soil O_2.

Wilted plants are more resistant to SO_2 than turgid plants (Zimmerman and Crocker 1934), a response attributable, at least in part, to stomatal condition. Soil water stress conditions affect SO_2 injury on beans and other species (Markowski et al. 1974). There is much less injury with drought conditions than with optimum soil water conditions. Sulphur dioxide increases the stomatal opening of drought-grown bean plants without affecting the high diffusive resistance and SO_2 protection afforded by drought conditions (Schramel 1975). Thus, in the case of SO_2, the water stress protection is not a function of stomatal closure. Humidity also affects SO_2 response. Plants at 30% relative humidity are one third as sensitive to SO_2 as those at 100% (Setterstrom and Zimmerman 1939). At high relative humidities, low concentrations of SO_2 (1 - 2 pphm) can induce stomatal opening even in the dark (Unsworth et al. 1972). This effect is particularly pronounced in plants grown under conditions of water stress. Transpiration would be increased as well as the uptake of other pollutants.

Atmospheric humidity affects F response of gladiolus (MacLean 1973) with more severe injury at 80% relative humidity than at 50% or 65%. Increasing irrigation frequency increases ethylene sensitivity of carnation flowers (Mayak and Kofranek 1976).

1.13 Nutrition

The effects of changes in mineral nutrition status have many complex aspects. Generally, mineral nutrient levels that promote growth have also been reported to increase pollutant injury to crops, but there are some notable exceptions. Specific responses to a particular pollutant must be considered on the basis of the particular nutrient element or elements and species under consideration.

Nitrogen. Nitrogen nutrition has been modified in oxidant studies with conflicting results. Exposure of spinach to peroxides derived from olefins results in greater injury to plants with an abundant N supply, than to plants grown under low or deficient N conditions (Middleton 1956). Oxidant injury to spinach and romaine leaves also is increased by additional N fertilization.

In contrast, the highest N treatment has the least oxidant injury on potato (Vitosh and Chase 1973). The speckle leaf O_3-injury symptom on potato occurs at the stage of maximum N uptake, with N-deficient leaves being most susceptible. 'Concord' grapevines also have less oxidant injury with added N compared to no added N (Kender and Shaulis 1976).

Increasing N level results in increasing injury on spinach due to ozonated hexene exposure (Brewer et al. 1962). There are significant interactions of P with N and with K. Addition of P alone decreases leaf production and oxidant injury. Addition of K without P does not affect leaf production, but usually increases the oxidant injury at low but not high N. When both P and K are added to spinach, however, severity of injury is significantly decreased. Essentially opposite are effects observed on mangels.

When radish plants are grown at different levels of N and P nutrition and different temperature regimes, temperature modifies the nutrient effect on O_3 response (Fig. 11). Growth suppression by O_3 treatment is more severe at high than at low N levels, particularly at cool temperature.

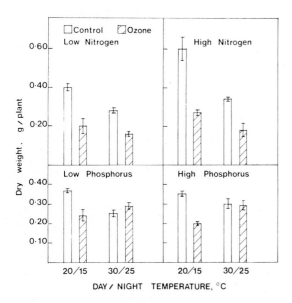

Fig. 11. Effects of growth temperature and of N and P nutrition on the response of radish plants to O_3 at 25 pphm for 4 hr. Means ± standard error (Ormrod et al. 1973).

Nitrate nutrition affects NO_2 response of beans (Srivastava et al. 1975a). Severity of injury is decreased by increasing nitrate supply. Nitrogen deficiency or supra-optimal N levels decrease SO_2 susceptibility of tomato (Leone and Brennan 1972a). Both deficiency and excess of N reduce F uptake from air or soil by tomatoes (Brennan et al. 1950). Nitrogen deficiency increases the HF sensitivity of beans (Adams and Sulzbach 1961).

Phosphorus. Increase in P supply to tomato plants results in increased O_3 injury (Leone and Brennan 1970), but P nutrition status has little effect on O_3 response of radish (Fig. 11). Phosphorus deficiency reduces F uptake by tomatoes (Brennan et al. 1950), has little effect on HF sensitivity of beans (Adams and Sulzbach 1961), but increases severity of F-induced tipburn of gladiolus (McCune et al. 1967).

Potassium. Injury due to O_3 on tomato foliage is reduced by K deficiency, which increases foliage diffusion resistance (Leone 1976). Foliar K content increases after O_3 exposure.

Potassium deficiency has little effect on HF sensitivity of bean (Adams and Sulzbach 1961) even though bean plants deficient in K take up more F than plants deficient in N or Ca (Applegate and Adams 1960a). Potassium deficiency increases severity of gladiolus F injury (McCune et al. 1967). Fluoride does not enhance K deficiency symptoms in tomato (MacLean et al. 1969).

Sulphur. A high plane of S nutrition (32 mg/l in nutrient solution) protects snap bean leaves from O_3 toxicity compared with plants at a low S nutrition level (1.3 mg/l) (Adedipe et al. 1972b). This protective action of high S against O_3 injury may be a result of elevated amounts of sulfhydryl (SH) groups in the plant tissues, since several chemicals containing SH groups can be used as protective sprays (Fairchild et al. 1959).

Sulphur nutrition effects on SO_2 sensitivity are an important consideration in relation to the mode of action of SO_2. Sulphur dioxide sensitivity of tomato plants increases with increasing S nutrition. The increased SO_2 susceptibility at high S levels is apparently a result of cumulative toxicity since increased injury parallels elevated foliar S absorption from both the nutrient solution and the SO_2 atmosphere (Leone and Brennan 1972a, b).

Calcium. Both Ca nutrition effects on F-toxicity and the use of F-protectant Ca-containing dusts and sprays have had special attention because of known Ca-F interactions. Deficiency or excess of Ca both reduce F uptake by tomatoes (Brennan et al. 1950). When Ca is added to the substrate in concentrations above 40 ppm, its tendency to precipitate F in the form of insoluble compounds within or around the roots further reduces the possibility of injury to the foliage. This action of Ca appears to be both physical and chemical. Calcium nutrition affects tomato response to HF with an F interference with Ca metabolism (Pack 1966). Calcium deficiencies have little effect on HF sensitivity of bean (Adams and Sulzbach 1961).

Lime foliar dusts or sprays protect gladiolus from F scorch, the Ca probably tying up the F in insoluble compounds on the leaf surface (Allmendinger et al. 1950). Foliar application of lime dusts or sprays also helps protect peaches from F-induced soft suture disorder (Benson 1959) with the mode of action the same as for F-protection in gladiolus. Sprays of $Ca(OH)_2$ can be used to reduce the effects of atmospheric fluorides on citrus trees and gladioli (Brewer et al. 1969a).

Magnesium. Deficiency of Mg increases severity of F injury on gladiolus but does not affect F accumulation (McCune et al. 1967). Fluoride enhances the apparent severity of Mg deficiency of tomato (MacLean et al. 1969). Deficiency of Mg increases F-injury (MacLean et al. 1976).

1.14 Other Environmental Factors

Air Velocity. There is a marked effect of air velocity on pollutant response as indicated by exposure of tomato and cucumber plants to either O_3 or SO_2 at two flow rates (Brennan and Leone 1968). The higher flow rate results in greater injury symptoms and greater uptake of S in the case of SO_2. Air velocity effects differ with species (Heagle et al. 1971). Bean injury increases with increasing air velocity but cucumber is unaffected.

Pollutant Pretreatment. Pretreatment with pollutants may alter subsequent response to the same pollutants. For example, bean plants pretreated with low O_3 have reduced injury from a later higher concentration O_3 exposure (Runeckles and Rosen 1974).

1.15 Protection Against Air Pollutants

Pollutant injury to horticultural plants may be reduced by the use of various chemical agents. A wide range of compounds have protective capability. Protective compounds may be grouped into antioxidants, which may counteract the effects of oxidant-type air pollutants; fungicides, which may provide protection against both fungal organisms and air pollution; growth regulators, which may have a dual function - as growth and development control agents and anti-pollutants; and other chemicals such as surface coatings which may physically protect plants or chemically react to detoxify air pollutants (Ormrod and Adedipe 1974).

1.16 Antioxidants

Numerous antioxidants have been tested as protectants against oxidant air pollutants. Ascorbic acid protects a wide range of crops, including bean, celery, romaine, lettuce, petunia and citrus, from leaf injury caused by oxidant-polluted air in the Los Angeles, California, U.S.A. area (Freebairn and Taylor 1960). Most species are partially protected, while a substantial protection in petunia is associated with increase in number and weight of leaves. Oxidant injury to bean plants sprayed with 0.01N K ascorbate is about 40% of injury to control plants. Both the K and Ca salts of ascorbic acid, when fed through the roots, also protect bean plants from O_3 injury (Freebairn 1963).

Numerous antioxidant chemicals in solution have been applied to cucumbers which were then exposed to O_3 (Siegel 1962). Prolonged contact with compounds like hydrazine, indole, tryptophan and mescaline results in effective O_3 protection, but ascorbic acid is ineffective in cucumber even though it is effective in bean and petunia (Freebairn and Taylor 1960). The low effectiveness of ascorbic acid on cucumber may be attributed to lack of any specific effect on O_3-induced growth alterations, or to possible auto-oxidation to dehydroascorbic acid. These reasons are largely negated by work with root uptake by beans (Freebairn 1963). The apparent conflict in these responses may be related to species differences in effective uptake of ascorbic acid. Ascorbic acid is less effective than nickel-N-dibutyl dithiocarbamate in protecting bean plants against oxidant injury (Dass and Weaver 1968).

Application of two antioxidants, DPPD and NBC, to young navel orange trees prevents some photochemical smog-induced fruit drop and results in greater fruit yields (Thompson and Taylor 1967). Under field conditions, manganous 1, 2-naphthoquinone-2-oxime protects tomato foliage from injury apparently caused by excessive atmospheric O_3 (Rich and Taylor 1960). The similar cobaltous and manganous chelates of 8-quinolinol are also effective antioxidants. These chemicals are applied at 405 mg/m^2 to cloth which is used in the same manner as for shade-grown tobacco. Compounds used as antioxidants in the rubber industry, for example, the dialkyl-p-phenylene diamines, are even more effective.

1.17 Fungicides

Fungicides may be air pollution protectants as a result of antioxidant or stomatal closing properties. In some cases, the mode of action is not known. Injury caused by exposure of plants to ozonated gasoline or hexene-1 is prevented by sprays or dusts of the fungicides zineb, maneb, thiram or ferbam (Kendrick et al. 1954). Such fungicides as Bordeaux mixture, dichlone and chloranil do not effectively protect bean plants against these pollutants. Field tests as well as laboratory experiments indicate that the degree of protection is directly related to concentration of these fungicide-chemical protectants (Kendrick et al. 1962). Action of the chemicals is local and not systemic, suggesting the deactivation of oxidants at the leaf surface.

The application of the fungicide PMA to leaves results in stomatal closure. Thus, PMA is a highly effective O_3 protectant (Fig. 12).

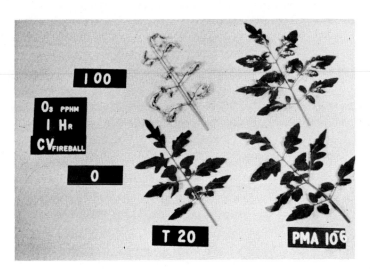

Fig. 12. Ozone injury to the fourth oldest leaf of tomato as influenced by 'Tween 20' surfactant spray and PMA spray containing the same surfactant (Adedipe et al. 1973).

Unrelated systemic fungicides, such as triarimol, benomyl and 1,4-oxathiins, also possess the ability to protect plants against O_3, when plants are grown in soil containing these compounds. Their ability to protect plants against O_3 is apparently not directly related to the mechanism of their effectiveness as systemic fungicides. Triarimol suppresses O_3 injury to greenhouse- and growth chamber-grown bean plants (Seem et al. 1972). Foliar sprays at a concentration of 50 mg/1 result in a four-fold reduction of injury. Protection is also obtained with as low as two mg/g soil applied as a drench. Two other systemic fungicides, thiophanate ethyl and thiophanate methyl, have also been used as chemical protectants against O_3 (Seem et al. 1973). Nearly complete protection from O_3 injury is achieved with 200 mg thiophanate ethyl/gm soil as a drench. Soil application of thiophanate methyl is not consistently effective, while foliar applications of both fungicides at the rate of 500 mg/l result in significant injury reduction. Thiophanate sprays also protect azaleas against oxidant injury (Moyer et al. 1974b).

Benomyl has been one of the most successful O_3 protectants. It has been widely tested on horticultural species and has provided significant oxidant protection qualities as well as being a useful fungicide. 'Pinto' beans are protected against O_3 injury by benzimidazole and benomyl incorporated in the soil (Pellissier et al. 1972a). Thiabendazole is ineffective, apparently because of lack of plant uptake from the soil. Benomyl soil drenches significantly reduce chronic O_3 injury in poinsettia (Manning et al. 1973b). Benomyl sprays reduce oxidant injury on grapevines if three or more applications are made during the season (Kender et al. 1973). The greater the frequency of application, the greater the protection. Benomyl or thiophanate provides a marked reduction in O_3 injury to annual bluegrass when applied as a soil amendment or drench (Moyer et al. 1974a). Benomyl soil drenches are an effective method of protecting azaleas against oxidant injury (Moyer et al. 1974b).

'Tempo' bean plants treated with benomyl have about a 30% higher yield than control plants (Manning and Feder 1976). This effect is attributed to protection against ambient O_3 injury. Benomyl or thiophanate ethyl applied biweekly, or ancymidol, a growth retardant, applied once, provide effective O_3 protection of chrysanthemum throughout the season. Ancymidol, however, has undesirable effects on flowering (Klingaman and Link 1975). The growth retardants triarimol and SADH are ineffective, as is an antitranspirant. Soil drenches of benomyl do not protect 'Pinto' beans against PAN or O_3 + PAN (Pell 1976), with the exception of older primary leaves on 20 day-old plants, when the benomyl is applied seven days before pollutant exposure.

Carboxin, another systemic fungicide, and other 1,4-oxathiin derivatives, protect beans and tomatoes against O_3 injury (Rich et al. 1974). Foliar diphenylamine spray or dust on apple or dust on bean, muskmelon and petunia provide protection against O_3 (Gilbert et al. 1975). Effective concentrations are about 1% as a dust and 500 - 1000 ppm as a spray. The fungicide diphenylamine as an oxidant protectant is also used to quantify ambient oxidant effects in monitoring air quality in Georgia, U.S.A. (Walker and Barlow 1974).

1.18 Growth Regulators

Auxin. Growth regulator interactions with air pollutants were studied first using oxidants and IAA (Hull et al. 1954). The studies were primarily concerned with measuring the effects of pollutants on the activity of IAA using Avena coleoptile bioassays. Inactivation of IAA by O_3 can be demonstrated using in vitro systems, but it is not known whether such an inactivation would be of significance in mature intact plants (Ordin and Probst 1962). Ozone effects on IAA have also been examined in short-term cucumber seedling experiments (Siegel 1962).

Growth Retardants. There have been a number of studies of the effectiveness of growth retardants as protectants against air pollutants. Just as in the case of fungicides, growth retardants could have the dual role in horticultural production of growth control and pollutant protection. The application of the growth retardant Phosphon D protects petunia against injury from irradiated automobile exhaust (Seidman et al. 1965). The growth retardants CBBP and SADH as foliar sprays are effective O_3 protectants on petunia (Cathey and Heggestad 1972). The effectiveness of these compounds in reducing injury is directly proportional to chemical dosage. The SADH protective effects are enhanced by combination with the antioxidant L-ascorbic acid and an antitranspirant coating (Folicote) in the spray. Several other growth retardants are ineffective in reducing O_3 injury of petunia. The growth retardants ancymidol and chlormequat reduce visible injury on poinsettia induced by either O_3 or SO_2 (Cathey and Heggestad 1973).

Other Growth Regulators. Other classes of growth regulators may also be effective protectants. Of several growth regulators, BA is the most effective as an O_3-protectant on radish (Adedipe and Ormrod 1972). Benzyl adenine prevents O_3-induced reductions of both leaf and root weights while GA and IAA protect against root weight loss only (Fig. 13). Application of CEPA is ineffective in preventing growth reductions. Abscisic acid application to the primary leaves of bean plants reduces O_3 injury to these leaves (Fletcher et al. 1972) (Fig. 14).

Anti-senescence compounds such as benzimidazole, N-6-benzyladenine and kinetin applied to bean plants protect them from visible O_3 injury (Tomlinson and Rich 1973). However, it should be noted that while kinetin or BA reduce O_3-induced chlorophyll loss in beans, they do not prevent weight loss induced by 10 pphm O_3 (Runeckles and Resh 1975b).

Protection against ethylene injury by growth regulators can also be demonstrated. Cytokinins, either kinetin or isopentyl adenine, combined with 5% sucrose, reduce ethylene sensitivity of carnations and increase longevity of flowers exposed to ethylene (Mayak and Kofranek 1976).

1.19 Other Chemicals

Other attempts to offset the adverse effects of O_3 exposures employ metabolites or anti-metabolites such as oxidase inhibitors used on tobacco

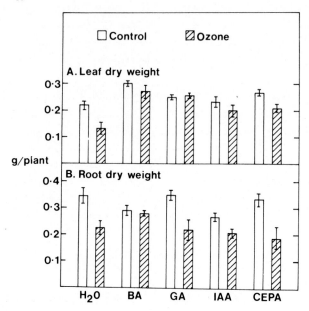

Fig. 13.   Effect of O_3 on (A) leaf and (B) root (radish) weights (Adedipe and
Ormrod 1972).

leaves (Koiwai and Kisake 1973). The free sterol content of bush bean leaves
decreases after three O_3 exposures (Tomlinson and Rich 1973). Treating leaves
with anti-senescence compounds such as kinetin, BA or benzimidazole results in
less O_3 injury as well as higher levels of free steroid in the leaves.

An antitranspirant is not an effective O_3-protectant on chrysanthemums
(Klingaman and Link 1975) but there has been some success with this type of
material. Water solution sprays of Folicote, a paraffinic hydrocarbon wax
emulsion, are effective O_3 protectants (Pellissier et al. 1972c). Only
sprayed areas of the leaf are protected.

Covering the leaf with powdered materials has been successful with tobacco
and may also protect horticultural species. Tobacco leaves treated with enough
dried particulate charcoal, diatomaceous earth, or powdered ferric oxide to
form a covering are relatively uninjured by O_3 (Jones 1963). The plants are
thought to be protected by these heterogeneous catalysts which when sprayed
or dusted on the plant, promote the decomposition of O_3 at a regulated distance
above the actual surface of the leaf.

Fig. 14. Bean leaves after exposure to O_3. The opposite primary leaves were treated with ABA or H_2O four hr before plant exposure to O_3 (Fletcher et al. 1972. Reproduced by permission of the National Research Council of Canada from the Canadian Journal of Botany, 50, pp. 2389-91).

A very successful method of chemical protection is the use of Ca applications to protect plants against F injury. For example, $Ca(OH)_2$ sprays improve citrus fruit production in an F pollution area (Brewer et al. 1969b) and lime dusts or sprays protect gladiolus (Allmendinger et al. 1950).

1.20 Beneficial Effects of Air Pollutants

Numerous examples can be cited of apparent desirable effects of air pollutants in terms of growth and yield enhancement. Unfortunately, the stimulating effects are usually the result of a particular combination of factors and many observations of beneficial responses do not seem to have general applicability. The literature on the effects of low concentrations of air pollutants has been surveyed and there are many reported cases of plant growth stimulation by pollutants compared to growth in filtered air, suggesting an adaptation by many species to ambient levels of pollutants (Bennett et al. 1974).

Oxidants. Ozone exposure results in stimulation of 'Pinto' bean lateral bud enlargement (Engle and Gabelman 1967). Auxin-like activity is found using Avena coleoptiles but a mechanism for this type of O_3 response has not been established. Ozone stimulates stem elongation within three days of exposure in a wide range of tomato cultivars, whether or not their leaves are sensitive in terms of visible injury (Neil et al. 1973). Ozone must therefore have some effect on the control system for stem enlargement.

There is enhanced nitrate formation and ultimately higher protein in pea leaves exposed to NO_2 (Zeevaart 1976). Nitrate and nitrite formed after NO_2 exposure are reduced to ammonia and amino acids, resulting in a higher leaf protein content. Leaf necrosis caused by NO_2 occurs at the same time in pea and other species. This injury response is attributed to acidification of tissue as a result of NO_2 and water reaction. Exposure to a mixture of NO_2 and NH_3 reduces leaf necrosis markedly. There is much more NO_2 injury in the dark than in the light when reducing power of the tissue is much greater.

Sulphur Dioxide. Possible beneficial effects of SO_2 may relate to the provision of nutritional S directly to the plant or via the soil. Sulphur is removed from soils with harvested crops and, just as for other nutrients, deficiencies will ultimately occur if there is no replacement. The problem is to know the level of SO_2 which is adequate for nutritional S supply but not injurious. Sources of SO_2 pollution seldom provide constant SO_2 concentrations in time and space so it would not be appropriate to assume, in most cases, that pollutant SO_2 can be utilized as a source of fertilizer S. Studies of the nutrient value of SO_2 show that plants grown under conditions of S or N deficiency are less susceptible to injury by high concentrations of SO_2 than is the case with plants growing without deficiencies (Leone and Brennan 1972a). Presumably plants deficient in N and/or S can tolerate higher ambient levels of SO_2 and might give higher yields in the presence of SO_2 than if under full nutrition.

Growth of white birch and pin oak seedlings responds to relatively high but subtoxic ambient SO_2 air (Roberts 1975). Sulphur dioxide has a beneficial effect on white birch, a sensitive species, in that the over all growth is greater in the high SO_2 environment while the growth of pin oak, an insensitive species, is best at lower SO_2 concentration. White birch apparently is able to metabolize and utilize the S from the surrounding air, possibly because stomates are not closed by SO_2.

Sulphur dioxide also has the potential beneficial effect of pathogen control. This gas affects not only expression of disease but also reproductive potential of pathogens. Concentrations of less than 20 pphm SO_2 decrease the incidence and severity of infection and the size and percentage of uredospores of bean rust (Weinstein et al. 1975). Sulphur dioxide acts as both a protectant and an eradicant although exposures are more effective before than after inoculation. Early blight of tomato is not affected under the same conditions. Black spot of roses is checked or eliminated by SO_2 (Saunders 1966).

Fluoride. Kalanchoe plants absorb F through roots in proportion to soil F content (Johnson and Applegate 1962). Height growth of plants is stimulated in direct proportion to F uptake. Fluoride does not stimulate plantlet formation on detached leaves.

1.21 Evaluation of Air Pollutant Effects

Field Studies. The use of field studies to evaluate air pollutant effects requires a knowledge of visible injury symptoms (Jacobson and Hill 1970). A knowledge of guidelines for field surveys and vegetation sampling is also necessary together with an appreciation of instrumental analyses and the use of bioindicators of pollutants (Weinstein and McCune 1970). Much of the early information on species sensitivity to oxidants was gained in the field in Southern California, U.S.A., where injury to sensitive vegetable crops and other horticultural plants occurs every year (Middleton et al. 1950). Similar field studies of air pollutant-induced injury to horticultural plants are now conducted in the vicinity of many urban and industrial areas throughout the world. Such field studies can also provide useful indications of injurious effects of other forms of pollution including that due to trace elements, salinity and pesticides.

Injury ratings are usually established as a result of visual estimates using numerical scales corresponding to descriptive terms or the proportion of surface area injured. Such visual rating systems are subject to bias as a result of lack of consistency from sample to sample and of variations among observers. There are various methods for reducing the effect of such observer bias including the use of more than one observer per sample (Todd and Arnold 1961). Another approach is to evaluate injury in terms of pigment destruction. The reduction of chlorophyll content of O_3-injured leaves is well correlated with visible injury, and a rapid procedure for chlorophyll analysis has been developed for determination of O_3 injury (Knudson et al. 1977). The use of reflectance spectrophotometry for the quantitative assessment of chronic O_3 injury has been evaluated (Runeckles and Resh 1975a). This method is considered even more precise than chlorophyll extraction and measurement.

A special type of field study is the use of selected sensitive plants as monitors of ambient pollution in specific regions. Certain plants grown under standardized management conditions can be used to give indications of natural levels of pollutants, to identify pollutants, and to indicate environmental effects on sensitivity. The principles of this technique have been established and the method is widely used (Fig. 15). Many horticultural plants are included

50

Fig. 15. Indicator plant plot including buckwheat for SO_2, 'Pinto' beans for
O_3 and PAN, gladiolus for F, and tobacco for O_3 (Author's photo).

in the species suggested as indicators of air pollution but there is a need for
better sensitive species. Those requiring little maintenance would be preferred.
For example, sensitive perennials could be planted at a test site and be used
thereafter while annuals must be planted each year. Also, the responses of
sensitive indicator species must be correlated more precisely with those of
horticultural plants of economic or aesthetic value in the area under study.

This use of plants as bioindicators of pollution and the need for a uniform
system of injury indices for such use have been discussed (Heck 1966). Requirements for adequate recognition of air pollution problems using indicator plants,
along with laboratory and greenhouse controlled exposure methods, have also
been summarized (Ormrod and Adedipe 1975). The full range of methods for monitoring air pollutants including both instrumental and low cost monitoring
devices has been described as well as the rationale for air quality monitoring
(Shenfeld 1975).

The methods used in the Netherlands for evaluating air pollution with
indicator plants are described by Van Raay (1969) who also discusses the

requirements for a successful monitoring programme. Both plant symptoms and
chemical analyses of plant tissue are used to trace sources, concentrations
and extent of air pollution. Endive, lucerne (alfalfa), clover, buckwheat and
barley are indicator plants used for SO_2, while gladiolus and freesias are used
for F detection. The need for both symptom observation and chemical analysis
is emphasized, especially for HF. The Netherlands monitoring system has been
more recently described in greater detail (Posthumus 1976). 'Snow Princess' or
'Flowersong' gladiolus are used for HF detection in summer and 'Blue Parrot' or
'Preludium' tulip in the spring. PAN is monitored by small nettle and annual
meadow grass (annual bluegrass), and O_3 by 'Bel W3' tobacco and 'Subito' or
'Dynamo' spinach. Ethylene injury is detected by 'White Joy' petunia and
'Bintje' potato, and SO_2 by 'Du Puits' lucerne (alfalfa) and buckwheat. Nitrogen dioxide is also detected by spinach and tobacco. Annual ryegrass is used
to detect trace element contamination. A uniform cultivation system is prac-
tised at all sites with all plants, with soil and containers prepared at one
site and transported to other sites on a frequent regular basis (Fig. 16, 17).
This indicator plant system is useful for the description of the seasonal
pattern of O_3 and PAN episodes (Floor and Posthumus 1977). Weekly injury
ratings are plotted for three years and four locations to illustrate patterns
of oxidant pollution in the Netherlands.

Fig. 16. Air pollutant monitoring at Wageningen, Netherlands. The soil boxes
          equipped with porous watering candles are placed in a water reservoir
          with tubes extending below the water surface and connected to the
          porous candles (Author's photo).

Fig. 17. Porous candles and tubes used in the automatic watering system for indicator plants used in the Netherlands (Author's photo).

A system for monitoring ambient air pollutants was also developed by Oshima (1974). A series of portable units are employed. Each unit is an enclosure with a reservoir and watering mechanism and is capable of sustaining several indicator plants for a week. Reference photographs are used to standardize injury assessments.

Pollen germination and tube elongation measurements are a possible method for quantifying plant injury from air pollutants. The technique has been used on lily pollen (Masaru et al. 1976) and petunia pollen (Feder et al. 1969). Another suggested measure of O_3 stress is the production of ethylene (Tingey et al. 1976).

Epiphytic bryophytes and lichens on balsam poplar can be used to assess and map the long-range effect of SO_2 pollution (LeBlanc et al. 1972). The number, frequency-coverage and relative resistance of the epiphytes are well associated with average ground level concentrations of SO_2.

A special type of field research to study oxidant effects uses the anti-oxidant compound 4,4-dioctyl diphenylamine applied to cheescloth to protect plants from O_3 injury (Walker and Barlow 1974). This approach permits the determination of effects of ambient O_3 at various locations on sensitive species such as petunia and tobacco. This general method could be used for other pollutants. Oxidant injury suppression by weekly benomyl sprays is used as a method of determining oxidant-induced bean yield reductions in the field (Manning et al. 1974). Oxidant injury suppression of up to 75 - 80% can be noted but only the most sensitive cultivars have a higher yield with benomyl treatment.

A method for exposure of tree branches to SO_2 in their natural environment has been developed (Spierings 1967b) (Fig. 18). The device is mounted so that a branch is exposed at the outlet. Transparent baffle plates are fixed on the sides of the outlet. Prevailing environmental conditions are considered to be relatively undisturbed.

Fig. 18. Apparatus for exposing branches of trees to pollutants (by courtesy of the Research Institute for Plant Protection, Wageningen, The Netherlands).

Chamber Studies. The use of chambers in which controlled concentrations of one or more pollutants can be maintained has demonstrated that field-observed injury can be duplicated for many sensitive species. Many chamber systems are in use, usually for a particular pollutant, but almost all chamber systems could be used for any gaseous pollutant or mixture of pollutants. Such facilities provide information on typical foliar symptoms and relative sensitivities of species and cultivars (Fig. 19).

Fig. 19. A simple O_3 exposure chamber mounted under fluorescent lamps in an air-conditioned laboratory. The fan in the small commercial O_3 generator also functions to move air past the plants (Author's photo).

More elaborate facilities are required for more detailed studies of pollutant-environmental factor interactions as well as for detailed physiological and biochemical studies. Environment control is necessary to allow separation of environment effects and to eliminate diurnal and seasonal variations.

Justification of the need for controlled environment facilities for pollution research has been provided by Hill (1967):

"The actual response of plants to an air pollutant can be markedly influenced by such environmental factors as light, relative humidity, and temperature during exposure to the pollutant. In studying the effects of air pollutants on plants, therefore, it is sometimes desirable to expose plants to the pollutants under carefully controlled environmental conditions. In this way, injury symptom expressions, concentrations required to cause injury, mechanisms of injury development, and suppression of growth can be studied with little interference from variable environmental factors. Without this control, results from different days may differ considerably."

Controlled light, temperature, humidity, air circulation and pollutant concentration are incorporated in the air pollution exposure facility constructed by Adams (1961). Another exposure chamber allows studies of O_3 phytotoxicity in the range of six to 100 pphm with control of temperature, relative humidity and air flow rate (Fig. 20). A fully controlled environ-

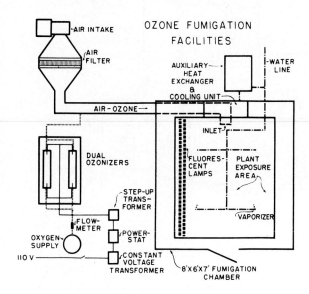

Fig. 20. Design and arrangement of a facility used for determining the effects of O_3 on plants (Menser and Heggestad 1964. Reproduced from CROP SCIENCE, 4, 103-5, by permission of the Crop Science Society of America).

mental chamber system in which wind velocity, CO_2 concentration, temperature, relative humidity, and pollutant concentration are controlled is in use for studies of several pollutants (Hill 1967). The chamber has an air-tight recirculating air system and allows measurement of net CO_2 assimilation and transpiration as well as pollutant uptake rate.

A commercial growth chamber can be modified for air pollution studies (Cantwell 1968) and such a chamber is being used for temperature and O_3 interaction studies (Fig. 21).

A system of chambers for growing plants in filtered air and exposing them to low concentrations of SO_2 has been described (Lockyer et al. 1976). Another unique exposure chamber was constructed for PAN studies (Fig. 22).

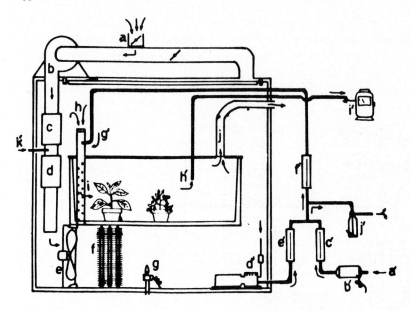

Fig. 21.   Controlled environment plant exposure chamber.   a. Air inlet to
conditioning chamber.   b. High pressure blower.   c. and d. Acti-
vated charcoal filters.   e. Circulating fan.   f. Refrigeration
unit.   g. Humidifying unit.   h. Inlet to exposure chamber.   i.
Distribution tube.   j. Air outlet tube. a^1. Oxygen outlet. b^1.
Ozone generator. c^1. Ozone flow controller. d^1. Pure air pump.
e^1. Pure air flow controller. f^1. Air pollutant flow controller.
g^1. Air pollutant inlet line. h^1. Pollutant sampling line. i^1.
Pollutant analyzer. j^1. Pollutant scrubber. k^1. Oxidant injec-
tion line (Cantwell 1968).

A plant growth and exposure system based on chemical reactor system
design has been developed (Rogers et al. 1977).   The concept of a continuous
stirred tank reactor (Fig. 23, 24) is utilized.   The design permits instan-
taneous mixing of incoming air with air already in the chamber.   Data are
obtained which can be used in mathematical modelling.

There are some difficulties involved in controlled environment pollutant
exposures (Runeckles 1974).   Attention should be drawn to the relationship of
experimental and naturally occurring concentrations, effective dosages, and
environmental conditions.

Fig. 22. Exposure chamber used for PAN studies by Posthumus at Wageningen, The Netherlands (By courtesy of the Research Institute for Plant Protection, Wageningen, The Netherlands).

Field Enclosures. A guiding principle for experimental exposures to pollutants is that field conditions must be approximated as far as possible if the purpose of the research is a critical evaluation of the pollutant effects on crop production. Abnormal environmental conditions should be avoided within enclosures if the implications of the research are to be applied to the natural habitat. Also, environmental conditions must be the same in separate enclosures used for different concurrent pollutant treatments.

Fig. 23. Exposure chambers of the dual continuous stirred tank reactor (CSTR) system (Photo courtesy of H. Rogers).

Fig. 24. Lower cart level: manifolds, rotameters, valves, stirring motors (Photo courtesy of H. Rogers).

Glasshouses are frequently used as field enclosures for pollution studies (Fig. 25). Details are available on the operation of glasshouses as F exposure

Fig. 25. Controlled exposure system utilizing glasshouses within a glasshouse
(By courtesy of the Research Institute for Plant Protection,
Wageningen, The Netherlands).

chambers (Hill et al. 1959). The same glasshouses could be used to study ambient pollutant effects by fitting control houses with air filters. A rotating exposure greenhouse with automatic environment controls and a system for maintaining SO_2 and HF concentrations has been described (Fig. 26).

A useful portable chamber system consists of Teflon film covered frames to which air cooling and O_3 generation equipment are attached (Fig. 27).

Two versions of a cylindrical open-top field exposure chamber are in use (Fig. 28, 29). The chambers provide an environment more closely resembling ambient conditions than the environment found in closed-top chambers. Temperature and humidity are similar to ambient, but light and rainfall distribution in the chambers differ somewhat from the open field. Chambers fitted with charcoal filters effectively protect sensitive plants from O_3 that severely injures plants outdoors or in unfiltered chambers.

Fig. 26. Four-chambered rotating exposure greenhouse (Matsushima and Brewer 1972)

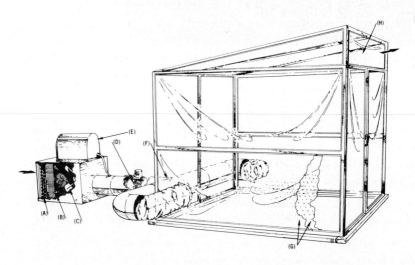

Fig. 27. Portable field exposure chamber system. A. Particulate filter. B. Activated-charcoal filter. C. Axial fan. D. Ultraviolet lamp. E. Mailbox used to house the ultraviolet lamp transformer and the variable autotransformer. F. Galvanized steel duct. G. Double-layered, clear Teflon duct with perforated inner layer. H. Perforated exit panel (Heagle et al. 1972).

In one version the chambers are fabricated from plastic film and filtered
or non-filtered air, or air with added pollutant, is blown through perforated
panels onto the plants (Fig. 28).

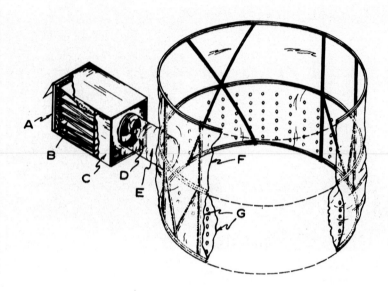

Fig. 28. Cylindrical, open-top field chamber: (A) fiberglas particulate
filter; (B) activated charcoal filter; (C) sheet metal box; (D)
0.5 hp axial blade fan; (E) connecting duct; (F) upper panel;
and perforated-lower-duct-panel (Heagle et al. 1973. Reproduced
from JOURNAL OF ENVIRONMENTAL QUALITY, 2, 365-8, by permission
of the American Society of Agronomy, Crop Science Society of
America and Soil Science Society of America).

In the other the chambers are fabricated from corrugated fiberglass panels
attached to aluminum hoops. A mobile air filter and blower assembly is
attached to an air delivery plenum within the perimeter of the chamber (Fig.
29). The efficiency of the filter systems has been tested along with the
distribution of introduced pollutant and, in the case of F studies, the
accumulation of F by plants. Additional information is available on the
characteristics of this open-top chamber for studies of the effects of air-
borne F and photochemical oxidants, including the relationship of chamber
environment to ambient conditions, contamination by ambient air, uniformity
of pollutant distribution, and realistic simulation of exposures (McCune et al.
1976). Such open-top chambers are now widely used in air pollution research

Fig. 29.   General view of cylindrical open-top chamber with mesh covering and
filter-blower assembly (Mandl et al. 1973.   Reproduced from JOURNAL
OF ENVIRONMENTAL QUALITY, 2, 371-6, by permission of the American
Society of Agronomy, Crop Science Society of America and Soil
Science Society of America).

Fig. 30.   Open-top exposure chambers at the Cambridge Research Station, Ontario,
Canada (Author's photo).

and there are numerous field installations (Fig. 30). They are also being used within glasshouses and in controlled environment rooms with irradiation provided by high intensity discharge lamps. A method has been developed to decrease mixing of ambient air into the open-top chambers during periods of high wind velocity (Kats et al. 1976). The installation of baffles around the top decreases the oxidant level inside a filtered chamber during windy conditions to about one-fourth that in filtered chambers without baffles.

Fluoride exposure poses special problems and information is available on the procedures involved in controlled F exposures of plants, along with the analytical procedures for F in plant tissue, water and air (Mavrodineau et al. 1962).

Ethylene treatment chambers covered with clear vinyl plastic are in use in a glasshouse (Fig 31). Other investigators use exposure chambers within a glasshouse for studies with various pollutants (Fig. 32).

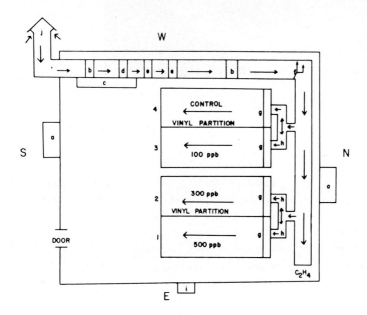

Fig. 31. Diagram of ethylene treatment chambers. a. Air-conditioning compressors. b. Fans. c. Control panel. d. Three-stage heating unit. e. Evaporators. f. Humidification unit. g. Diffusion plants. h. Point of C_2H_4 injection. i. Greenhouse exhaust fan. j. Air intake (Piersol and Hanan 1975).

Fig. 32. Exposure chambers within a glasshouse at Riverside, California, U.S.A.
(Author's photo). These chambers have a single-pass flow system and
were developed at the National Center for Air Pollution Control,
Cincinnati, Ohio, U.S.A. (Heck et al. 1968).

Pollutant effects on CO_2 exchange rates of intact leaves are measured
using systems such as that described by Taylor et al. (1965) for use on citrus
trees in the field. The system can be used to study the effects of either
ambient pollutants or controlled concentrations of pollutants. An installation
for exposure of plants to very low concentrations of H_2S under defined environmental conditions has been developed (Oliva and Steubing 1976).

CHAPTER 2

TRACE ELEMENT POLLUTION

When present in high enough concentrations trace elements may be toxic to plants. Many of the most common trace element contaminants in the environment are those widely involved in industrial processing and manufacturing. Such elements include aluminum (Al), arsenic (As), boron (B), cadmium (Cd), chromium (Cr), copper (Cu), lead (Pb), mercury (Hg), molybdenum (Mo), nickel (Ni) and zinc (Zn). Among the elements most likely to exceed the upper limits of acceptability are B, Cd and Pb.

The mineral elements have numerous roles and there are many consequences of excesses (Cottenie et al. 1976). There are differences in uptake patterns among elements as well as a number of factors affecting availability. The sources of toxic amounts of trace elements in soils are mainly fall-out in industrial areas which leads to accumulation over time, dumping of industrial waste, use of sewage sludge in agriculture, intensive fertilization with some types of manure, infiltration of contaminated waste water, and use of some pesticides. Plants may adapt to mineral stress including that due to trace element toxicities (Wright 1976). Screening methods have been developed for determining species adaptability.

2.1 Trace Elements of Concern

Aluminum. Studies of Al toxicity are largely based on providing excesses in nutrient solutions. Aluminum toxicity to peach seedlings in sand culture is manifested by severe restrictions of root growth as a result of collapse of epidermal cells (Kirkpatrick et al. 1975). Basal constriction of root hairs occurs because root hair eruption sites fail to heal properly in the presence of Al. Increasing Al concentration in nutrient solution results in marginal chlorosis of peach seedling leaves and growth of irregular shaped roots (Edwards et al. 1976). The foliar toxicity symptoms of midrib collapse, terminal dieback and defoliation resemble Ca deficiency. Aluminum in nutrient solutions decreases diffusive resistance of peach seedling leaves with increasing concentration (Horton and Edwards 1976). Decreased resistance appears to be more closely related to root volume than to stomatal aperture or density.

Arsenic. There is little information on the role of As as a toxic trace element in horticultural production. Several investigators have undertaken As analysis of soil samples from areas in which As had been used as a pesticide and there is documentation on phytotoxic As residues on vegetable farms (Jacobs et al. 1970), on As accumulation in agricultural soils of Ontario, Canada (Miles 1968), and on phytotoxic residues of As in orchard soils in Nova Scotia, Canada (Bishop and Chisholm 1962).

Soil As residues in former potato fields are associated with scorch and premature casting of needles and death of fine roots of hemlock in nursery plantings (Sinclair et al. 1975). Intensity of symptoms varies directly with As content of tissues and Cu concentration is not related to injury. Such a problem may arise when ornamentals nurseries move onto land formerly used for potato production. The As apparently comes from insect control chemicals and vine killers used in potato production on the same land in earlier years. In one case, an estimated average of 8.7 kg As and 24 kg Cu per ha had been applied per year for at least 20 years (Sinclair et al. 1975). Leaching rates of As are very slow indicating that As-sensitive species should not be grown on such contaminated soils.

Arsenic emitted from secondary Pb smelters does not result in above-normal As concentration in the edible parts of vegetables grown near the smelters even though the soils and other vegetation are contaminated (Temple et al. 1977). About 60% of the soil As is not available in As-contaminated field soils (Benson 1976). Tree growth is not well correlated with soil As suggesting that other factors have more influence in growth retardation than soil As. Apple tree roots growing in As-contaminated soil have sparse mycorrhiza development as well as stunted growth (Trappe et al. 1973).

Cadmium. There have been many studies of Cd toxicity in relation to horticultural crop production. The principal concern about Cd is that its accumulation in man is associated with cardiovascular and hypertensive diseases (Fassett 1975). Cadmium pollution is predominantly attributed to metal processing, impurities in Zn base oil additives, and wear of vehicle tires in which Cd is an impurity in Zn-containing additives. Some pesticides and superphosphate fertilizers are also sources of Cd. Cadmium accumulation in plants results in direct effects in terms of growth retardation and lesion development. Its presence in plant leaves may increase O_3-induced injury (Czuba and Ormrod 1974). This element may be found in the air, in solution, or in the soil. Plant responses depend on soil pH and liming, soil organic matter and cation exchange capacity, soil moisture, oxidation-reduction conditions, and species.

Superphosphate fertilizer may be a significant source of Cd in many vegetables (Schroeder and Balassa 1963). Phosphatic fertilizers manufactured in Australia contain 18 to 91 ppm Cd (Williams and David 1973). The addition of phosphate or its Cd equivalent as $CdCl_2$ to soil results in increased Cd content of the edible portion of vegetables. Plant uptake of the Cd is small, however, and ranges from 0.4 to 7% of that available. Processed green peas contain only up to 0.015 ppm Cd.

Nutrient solution studies reveal the response of vegetable species to $CdCl_2$ (Turner 1973). At one ppm Cd in the nutrient solution, Cd concentration in carrot tops is 2.2 ppm and in tomato 158 ppm. Cadmium treatment also frequently results in increased uptake of Zn in plant tops (Table 9). The wide

Table 9. Yield and Cd and Zn accumulated in tops of plants treated with a range of Cd concentrations in nutrient solutions.

Plant species	Cd treatment	Yield	Cd in tops		Zn in tops		Zn/Cd conc. Ratio
			Conc.	Uptake	Conc.	Uptake	
		g	g/g	g/plant	g/g	g/plant	
Beet	0	13.8	0.5	6.9	6.9	93.8	14
	0.01	14.4	0.9	13.0	7.0	100.8	8
	0.10	14.2	0.7	10.0	9.0	127.8	13
	1.00	6.4	2.5	16.0	13.2	84.5	5
Carrot	0	7.3	0.2	1.5	11.4	83.2	57
	0.01	5.5	1.5	8.3	14.7	80.9	10
	0.10	5.2	0.2	1.0	13.8	71.8	69
	1.00	2.0	2.2	4.4	23.0	46.0	10
Radish	0	10.8	0.3	3.2	13.7	148.0	46
	0.01	33.2	0.7	23.2	9.3	308.8	13
	0.10	27.9	14.9	415.7	15.1	421.3	1
	1.00	20.0	144.2	2884.0	21.5	430.0	0.2
Tomato	0	17.3	0.3	5.2	20.5	354.7	68
	0.01	17.7	1.4	24.8	24.3	430.1	17
	0.10	9.6	21.5	206.4	27.3	262.1	1
	1.00	8.0	158.0	1264.0	32.6	260.8	0.2

Data from Turner (1973). Reproduced from JOURNAL OF ENVIRONMENTAL QUALITY, 2, 118-9, by permission of the American Society of Agronomy, Crop Science Society of America and Soil Science Society of America.

differences among species suggest that one way to cope with high Cd soils would be to vary the plant species. Studies of the sensitivity of several vegetable crops to Cd and the relationship of tissue Cd concentration to solution Cd, indicate wide differences among species (Page et al. 1972). Increased K supply or higher nutrient solution pH reduce uptake and translocation of Cd in lettuce plants (John 1976). Plant Cd is also reduced, in some tissues, by additional Ca, Zn, P or Al. Solution Cd variation results in effects on tissue P, Fe, Mn, Al and Ca.

Soil factors affect Cd concentration in plants (Haghiri 1974). Cadmium concentration is decreased by increasing cation exchange capacity or organic matter in the soil, and increased by increasing soil temperature or adding Zn in the five to 50 ppm range. The increased Cd concentration is primarily due to reduced plant growth.

The relationship of Cd in plants to soil factors is clarified by the results of a study in which thirty soils were treated with Cd and seeded to radish and lettuce (John et al. 1972). Cadmium is taken up by these species in amounts related to acetate-extractable Cd in the soil, soil Cd adsorption, soil pH and organic matter. Plant growth is reduced by Cd and chlorotic symptoms appear (Fig. 33). Radish shoots accumulate more Cd from Cd-treated soil than roots,

Fig. 33. Compared to control plants, radish plants on the left show stunted growth and chlorosis of leaves due to the addition of Cd to a silt loam soil (John et al. 1972. Reprinted with permission from Environ. Sci. Technol., 6, 1005-9. Copyright by the American Chemical Society).

and lime addition to the soil decreases Cd uptake (Table 10). Comparisons of Cd in various plant parts of several species grown in Cd-enriched soil show a wide concentration range with generally most Cd in roots, followed by leaves, then fruits (Fig. 34). Cadmium concentrations tend to be lower in the edible seed, fruits, or tubers than in the roots or leaves of plants grown in soil with added Cd (MacLean 1976). Increased organic matter content decreases Cd

Table 10. Cadmium content (mg Cd/kg of oven-dry material) of radish parts as functions of soil-applied Cd and lime.

Cd added (mg/kg)	Tops		Roots	
	Unlimed	Limed soil	Unlimed	Limed soil
0	11.0 ab^z	9.5 a	5.1 a	5.4 a
0.5	13.1 ab	14.3 ab	8.0 a	6.3 a
1	12.8 ab	11.9 ab	6.5 a	6.2 a
5	36.6 b	30.4 ab	9.8 ab	8.3 a
25	125.6 d	70.7 c	34.5 c	18.8 abc
50	227.2 f	105.5 d	53.1 d	25.6 bc
100	402.8 g	183.1 e	174.3 e	28.3 c

^{z} Mean separation within tops or roots by Duncan's test, 5% level.

Data from John (1972b)

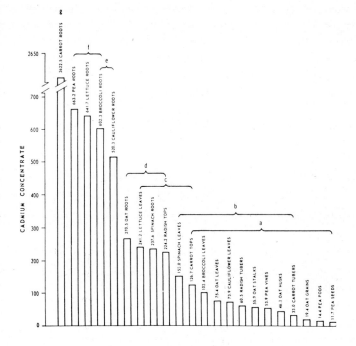

Fig. 34. Cd in various plant parts (John 1973).

uptake as does liming of some acid soils.

Cadmium affects photosynthesis and transpiration of corn leaves (Bazzaz et al. 1974b). Reduction in both processes is directly dependent on Cd concentration and becomes more pronounced with time. The two processes are highly correlated suggesting that Cd induces stomatal closure.

Ways of reducing dietary Cd uptake are of concern and one approach is to evaluate cultivar differences in Cd uptake. There are differences in yield and leaf Cd concentration among lettuce cultivars, providing evidence that Cd uptake and translocation is under genetic control (John and Van Laerhoven 1976a). Cultivars may be selected that have minimal Cd in the edible plant parts.

Chromium. There has been little study of Cr effects on horticultural plants. This element appears to be relatively innocuous under soil conditions. Application of Cr to a soil in which snap beans were growing showed that yields are reduced only if EDTA is also added (Wallace et al. 1976). In solution culture 10^{-5} M Cr is toxic with leaf Cr levels about 30 ppm. Chromium toxicity decreases cations in the plant.

Copper. Studies of Cu toxicity relate mostly to the use of Cu as a pesticide. Copper compounds have been used for many years as fungicides and bactericides and Cu sprays are still recommended for disease control on many vegetable crops (Walsh et al. 1972). In the past, Bordeaux mixture ($CuSO_4$ mixed with lime) was used for plant pathogen control while $Cu(OH)_2$ has been widely used to combat diseases. Copper toxicity from such pesticide use could develop in sandy soils because of their low cation exchange capacity and organic matter content. The effects on yield of the accumulation of Cu applied for pathogen control is of concern (Walsh et al. 1972). Slight yield decreases of snap beans are noted when the rate of Cu exceeds 130 kg/ha with a marked reduction at 405 kg/ha of Cu as $Cu(OH)_2$ or 486 kg/ha of Cu as $CuSO_4$. Toxicity is not ameliorated over a two year period.

Plants within 20 m of Cu high voltage electric transmission lines have higher Cu as a result of corrosion of the cables (Kraal and Ernst 1976). Greatest accumulation of Cu in this situation is by plants growing in sandy soil.

The mode of action of Cu toxicity to corn roots has been studied using complete nutrient solution containing eight mg/l Cu^{++} (Hunter and Welkie 1977). Post treatment with IAA, niacinamide, thiamin or sucrose has no effect on subsequent growth but excess KCl and succinic acid-2, 2-dimethyl hydrazide in the Cu solution double average growth and increase recovery. Rinsing with EDTA decreases Cu injury with 100% recovery if done within one hr of Cu treatment. Increasing pH increases Cu toxicity. An initial phase of Cu stress is sensitive to amelioration procedures but after three to six hr there is irreversible cessation of growth.

A Cu-tolerant and a non-tolerant strain of Mimulus guttatus have been used

to study the importance of pH in heavy metal toxicity (Soliman 1976). The
tolerant strain is adapted to acidic pH, while the non-tolerant grows better
at neutral pH. The F_1 hybrid is intermediate between parents at the lower pH
range but overdominant at pH 6.7, an indication of environment (pH)-dependent
heterosis.

Lead. Numerous studies of Pb effects on horticultural plants have been
conducted with the emphasis placed on the effects of Pb accumulation near roadways and industrial plants as well as on methods of ameliorating Pb toxicity.
The origins of pollutant Pb are not clear-cut and some studies have been conducted
to determine the relative importance of various Pb sources. For example, perennial ryegrass and radishes were used to establish that sources of leaf tissue Pb
are both the soil and the air, but not Pb in simulated rainfall (Dedolph et al.
1970; Ter Haar et al. 1969). The soil contribution is relatively fixed while
the balance in leaves is quantitively related to atmospheric Pb concentrations.
Radish root Pb is less than that in leaves and relatively unaffected by external
Pb concentrations.

Traffic density affects Pb concentration in soil and vegetation along road-
sides (Goldsmith et al. 1976; Motto et al. 1970; Page and Ganje 1970). Greatest
soil Pb levels are associated with highest traffic volume and decrease significantly with increasing distance from the roadway. No Pb accumulates with fewer
than about 30 vehicles per square km. Surface Pb is two to three times higher
than normal with a density of greater than about 200 motor vehicles per square
km. The soil Pb content does not exceed 52 ppm, considered to be below the
toxic level for plants, in surface soil from areas with high traffic density.
Vegetation Pb concentrations have the same spatial pattern as the soil contents.
Much of the plant Pb is from removable surface contamination. Plants obtain Pb
through leaves and roots with little translocation and the lowest Pb concentrations are in the flowers and fruit, or in the edible portion of vegetables such
as carrots, corn, potatoes and tomatoes. Vegetables within eight m of a road
may have 80 to 115 ppm Pb (Cannon and Bowles 1962).

Vegetables and fruit accumulate Pb, when grown in the vicinity of a Pb
smelting plant, to concentrations considered to be a potential health hazard for
consumers (Auermann et al. 1976). Tissue concentration of Pb may be 10 to 100
times higher than normal levels. Vegetables and fruit growing in such areas
should be abandoned and it is suggested that Pb-contaminated soils be physically
replaced with non-contaminated soils because of the long-term nature of Pb contamination.

Corn growth and yield are unaffected by Pb added to field soil (Baumhardt
and Welch 1972). Kernel Pb content is not affected by Pb added to the soil

although leaf and stem contents are increased to some extent. Attempts to determine the location of accumulated Pb in corn tissue indicate that surface Pb precipitate forms first on the roots, then Pb slowly accumulates in the cells in dictyosome vesicles (Malone et al. 1974). These vesicles migrate to the cell walls and the Pb ultimately accumulates in the cell wall outside the plasmalemma. Lead accumulates similarly in stems and leaves.

Uptake of Pb by bean and corn roots is not influenced by the presence of 2,4-dinitrophenol, sodium azide or low temperature but is dependent on solution pH (Arvik and Zimdahl 1974b). Uptake is reduced in bean but increased in corn with increased pH from 5.5 to 7.0. Only extremely small amounts of Pb penetrate the cuticles of leaves and fruits (Arvik and Zimdahl 1974a). Wax removal may increase Pb penetration but foliar uptake is not well related to cuticle thickness among species. Lead bromochloride aerosols also do not penetrate cuticle; instead they remain as a topical coating on the foliage.

Lead uptake by corn and sunflower is directly proportional to transpiration rates with reductions of photosynthesis and transpiration directly related to Pb concentration (Bazzaz et al. 1975). Stomatal resistance is apparently increased by Pb. Threshold tissue Pb concentrations for effects on transpiration and photosynthesis are 60 ppm in loblolly pine and 72 ppm in autumn olive (Rolfe and Bazzaz 1975).

Lead sources, lime and N have effects on Pb uptake by lettuce (John and Van Laerhoven 1972). Soil application of Pb increases Pb uptake by lettuce, and lime application represses Pb uptake. Application of N reduces Pb uptake by roots only. There is another approach to reducing plant uptake of Pb (Isermann 1977). Shoots are washed with AcEDTA chelate solutions or Na-polyphosphate solutions to reduce contamination and uptake of Pb by plants in areas where airborne Pb may be a problem. The chelated Pb is taken up much less readily resulting in restricted uptake by plants. The Pb status of old orchard land should be assessed before planting some vegetable crops, as well as the Pb and As response of the crops to be grown (Chisholm 1972). Lead levels in some vegetables grown on such soils exceed residue tolerances.

Many facets of Pb behavior in soils and vegetation, and Pb uptake and distribution remain to be elucidated (Zimdahl 1976). The only consistent generalizations are that increasing soil Pb availability increases plant uptake, uptake decreases with increasing soil P, organic matter content and pH, and large amounts of Pb are deposited on the foliage but most remain on the surface. The lack of toxic effects of Pb on plants is attributed to immobilization in vesicles and deposition in cell walls.

Mercury. This element enters the atmosphere from coal burning and smelting of Cu, Pb and Zn ores. Mercury settles with particulate matter to contaminate soil and plants. It is also added to the plant-soil system by its use in herbicides, in seed disinfection and in plant disease control. Mercury readily injures plants, causing growth reduction and toxicity symptom development.

The application of $HgCl_2$ to soil has formed the basis for an understanding of responses of several vegetable crops to Hg (John 1972a). Roots generally have the highest Hg content and, among edible parts, spinach leaves and radish roots have the highest Hg and pea seeds have the lowest Hg concentration (Table 11).

Table 11. Influence of Hg in soil on Hg in various parts of seven food crops.

	Plant part	Plant Hg (g/gx10^3)		
		Control	4 gHg/g[y]	20 gHg/g
Leaf lettuce	leaves	31 a[z]	17 a	45 a
	roots	112 a	175 a	387 a
Spinach	leaves	94 a	339 b	695 c
	roots	95 a	1022 b	1067 b
Broccoli	leaves	63 a	78 a	29 a
	roots	171 a	505 b	1870 c
Cauliflower	leaves	79 a	68 a	61 a
	roots	19 a	197 a	2447 b
Peas	seeds	1 a	2 a	3 a
	pods	5 a	11 a	42 a
	vines	110 b	187 c	85 a
	roots	11 a	160 a	1415 b
Radishes	tops	237 a	218 a	585 a
	"tubers"	13 a	26 a	663 b
Carrots	tops	24 a	61 b	72 b
	"tubers"	44 a	53 a	39 a
	roots	163 a	428 a	1058 b

[y] Hg concentration mixed in soil

[z] Mean separation within rows by Duncan's multiple range test, 5% level. Data from John (1972a).

Hg is translocated from apple leaves sprayed with PMA to developing fruit and to new terminal growth (Ross and Stewart 1962). Other Hg compounds sprayed on apple trees similarly result in Hg in the fruit (Stewart and Ross 1967). Mercury is also translocated from potato foliage sprayed with MCO to the tubers with most rapid accumulation during the most rapid sizing period (Ross and

Stewart 1964). Evaluation of soil Hg residues in an apple orchard indicates that they are almost entirely located in the top five cm of soil.

Cucumber root growth is inhibited by $HgCl_2$ and the shoot and root are disoriented (Puerner and Siegel 1972). Tests of other elements showed that disorientation is also induced by Zn, Ag, Cd, In and Pt, but not by Pb and Au.

Molybdenum. The application of sewage sludge to fields may result in the production of Mo-rich plants, which cause metabolic disorders in animals (Lahann 1976). The Mo concentration in sewage sludges ranges from two to 1000 ppm in the U.S.A. and Canada. Molybdenum is relatively mobile in the neutral to alkaline pH associated with sludge-amended soils, while other elements such as Zn, Cu, Ni and Cd are relatively immobile.

Tin. Little attention has been given to Sn effects on horticultural species. Little Sn is translocated to bean shoots from roots in either soil or nutrient solutions (Romney et al. 1975). Toxicity in 10^{-3}M $SnCl_2$ solution is prevented by the addition of $CaCO_3$. Tin is concluded to be relatively unavailable to plants and not readily mobile in them.

Trace Element Mixtures. As for air pollutants, the field situation is likely to be the existence of mixtures of potentially phytotoxic trace elements. Data on trace element content of horticultural soils are now available to illustrate this. For example, data have been gathered on the long-term effects of accumulated metal additions to soils in Ontario, Canada. Soils from 300 farm fields of known use history were sampled and analyzed for 11 elements. The metal additions, as a part of normal management practices, took the form of pesticides, trace elements, fertilizers, feed additives and industrial effluents. Amounts accumulated ranged widely according to the history of metal usage or presence of industrial sources. Pesticides contributed As and Pb to orchard soils, and As to potato fields, while fertilizers and manures made little contribution to trace element composition. Sludges contributed to high soil Cd in a few cases.

Vegetation and soil analyses in a Wales valley reveal significant levels of Pb, Cd, Cu and Ni (Burton and John 1977). These trace elements apparently result from aerial deposition, with motor vehicles the main source of Pb, Cd and Ni. Metal content is correlated with traffic frequency and urbanization. Soil Cd, Ni, Pb and Zn decrease with distance from traffic and with depth in the soil profile according to analyses conducted by Lagerwerff and Specht (1970). The contamination is associated with deposition of residues of gasoline, motor oil and tires in the roadside areas.

Both leafy vegetables and soil in the vicinity of a Cu smelter have elevated

levels of the trace elements Cu, Zn, Pb, Cd, Ni and Fe (Beavington 1975). Metal
contents are associated with distance from the smelter; soil and plant levels of
Cu and Cd were well correlated while Zn, Pb and Ni were not. Lettuce accumulates
more of most metals than do other species. There is concern about the possible
health hazards involved in consumption of such metal-containing vegetables.

Experimental modification of trace element nutrition allows study of interactions. For example, exposure of American sycamore seedlings to soil treated
with Pb, Cd or Pb plus Cd, permits studies of trace metal interactions (Carlson
and Bazzaz 1977). Root growth, stem diameter increment, and stem and leaf
growth are synergistically decreased by Pb plus Cd treatment while reductions in
photosynthesis and transpiration are not synergistic. An interaction study
of Pb and Cd in corn plants revealed a tendency for increased soil Pb to
increase plant Cd concentration and total uptake (Miller et al. 1977). Increased
soil Cd reduces Pb uptake. Both elements reduce vegetative growth of corn shoots
and there is a positive interaction, that is, the effects of the combined elements
are more deleterious than the additive effects of the separate elements. Combinations of Pb and Cd have significantly greater toxic effects than Cd alone which
is much more toxic than Pb alone to radical elongation of germinating corn seeds.
The effects of the combination are greater than additive, at least in part due
to elevated accumulation in combination treatments.

Comparative Studies with Trace Elements. Trace element toxicity levels in
tissue culture media for several vegetable crops have been established (Barker
1972). For Pb the toxicity threshold is somewhere between 0.5 and 5.0 mg/liter;
Hg toxicity occurs at 0.005 to 5.0 mg/liter depending on species. The toxic
range for Cu is 5.0 to 50.0 mg/liter and the toxicity level for Zn ranges widely
among species. Application of Cd, Co, Cr, Cu, Mn, Mo, Ni, Pb and V to several
species in sand cultures allowed study of trace metal element toxicities (Hewitt
1953). Copper always induces typical Fe deficiency with differing crop susceptibility. Nickel induces symptoms resembling Mn deficiency in potato and tomato.
Cobalt toxicity in tomato initially resembles Mn deficiency. Cobalt and Ni
responses vary greatly with species.

Cadmium, Pb or Zn additions to soil are reflected in only small changes in
radish tissue content (Lagerwerff 1971). Increasing soil pH from 5.9 to 7.2
decreases metal uptake. Aerial contamination accounts for 40% of the metal
content near a busy highway but only Zn is translocated significantly to the
roots.

Sewage Sludges. There has been considerable interest in the use of sewage
sludge as a source of nutrients and organic matter in horticultural crop production. Also horticultural areas might be readily available for disposal of these

waste materials. Sewage sludges are considered a source of plant nutrients but it is not well known how to utilize them effectively. Sludges are usually low in N and P so application of sufficient sludge to meet N and P needs may result in excessive trace elements. Sludges also may be used as organic soil amendments. The trace elements may not accumulate to phytotoxic levels if sludge is used in moderate amounts. However, a major concern is the potential danger to consumers of food crops, especially if trace elements accumulate in the edible portion of plants. Acceptable use of sewage sludges will depend on proper selection of crop species, soil type and sludge composition. Sludges high in trace elements could be harmful to both plants and animals if disposed of on food-producing land.

Many research projects have indicated that severe problems occur when sewage sludges are applied to horticultural crops. For example, sewage sludge was mixed with several soils at 0, 10 or 25% by volume and resultant tomato growth observed in the greenhouse (Touchton and Boswell 1975). Addition of any sewage sludge severely inhibits growth, causes chlorosis, and even results in death of some plants (Table 12). The injury is considered due to excess Zn in the soil and

Table 12. Growth response of tomato plants to sewage sludge applications.

Sewage sludge rate (%)		Soil			
		Davidson	Lynchburg	Montevallo	Norfolk
Height (cm)	0	19.0 a[z]	20.0 a	19.0 a	13.2 b
	10	6.8 c	5.9 cd	5.6 cd	6.1 cd
	25	5.0 cd	5.2 cd	4.4 d	4.9 cd
Green weight (g)	0	19.0 a	20.0 a	19.0 a	9.0 b
	10	1.8 c	1.0 c	0.5 c	1.1 c
	25	0.6 c	0.6 c	0.4 c	0.6 c
Dry weight (g)	0	2.82 a	2.63 a	2.53 a	1.05 b
	10	0.21 c	0.12 c	0.07 c	0.14 c
	25	0.08 c	0.08 c	0.04 c	0.08 c
Dry weight (g)/ height (cm)	0	0.14 a	0.13 a	0.14 a	0.08 b
	10	0.03 c	0.02 c	0.01 c	0.02 c
	25	0.02 c	0.02 c	0.01 c	0.02 c

[z] Mean separation within a factor by Duncan's multiple range test, 5% level. Data from Touchton and Boswell (1975).

plant, with sludge treatment resulting in about 25 times more Zn in plants than in controls. Soil type has little effect on tomato response to sludge application. Elevated levels of Mn, Cu, Cd, Cr and Pb may also contribute to the growth retardation and chlorosis.

Plant responses to sewage sludge application are complex according to experiments on the use of sludges and lime on lettuce and beets (John and Van Laerhoven 1976b). Tissue metal content is dependent on the sludge metal content and also on the type and application rate of sludge, lime regime, and release from organic matter. Sludges high in Zn, Cu, Cr or Ni, mixed with limed soil were used to study yield and tissue mineral composition of a corn-rye-corn succession (Cunningham et al. 1975a). High soluble salts from high sludge application rates result in reduced yield of the first corn crop. Addition of N, P and K by the sludge stimulates yield at lower sludge application rates. Tissue metal concentrations increase in proportion to sludge application rate. Addition of Cu and Zn to applied sludge decreases corn yield, while Ni has little effect, and Cr addition increases growth (Cunningham et al. 1975b, c). Toxic tissue concentrations of Cu are greater than 20 ppm and Zn greater than 400 ppm. Increasing Cr is associated with decreased tissue concentrations of other metals. There are Cu-Zn interactions and Cd concentration increases with increasing Cu.

Sludge applications each year to fields continuously planted to corn reveal some interesting effects (Hinesly et al. 1977). The assimilation of Zn and Cd is determined by the sludge-borne Zn and Cd applied immediately prior to planting rather than by the total applied in all years. If sludge applications were terminated Cd content of corn kernels would thus be expected to decrease to normal levels. Addition of Cr, Cu, Zn or Ni to soil compared with the same additions as sewage sludge reveals important differences (Cunningham et al. 1975c). Addition of elements to soil results in lower yields and generally higher metal concentrations than equivalent sludge treatments. Inorganic salt treatments are thus not useful for evaluation of phytotoxicity and metal uptake from sewage sludges.

Mixing sewage sludge with acid soil limed to give a pH range of 4.3 to 6.8 illustrates the interaction of metals and soil pH (Bolton 1975). Ryegrass yields are reduced by sludge at pH less than 6.0. Increasing sludge amounts results in sludge pH effects on soil pH. Sludges to be used on agricultural land should be adjusted to pH 7 to minimize possible heavy metal injury to crops.

The use of "chemical" sludges on tulips has been evaluated (Kirkham 1977a). The material, consisting of ferric chloride-lime treated dewatered sludge, is added to a greenhouse rooting medium. Growth is stunted compared to plants grown with dried organic sludge, for which growth is less than with liquid organic sludge, effluent, and water (Fig. 35). Lime added to condition sludge at a sewage treatment plant reduces availability of the trace elements as well as K and P.

78

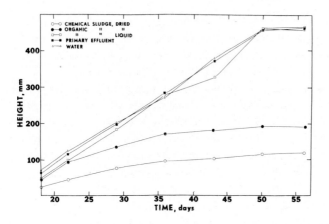

Fig. 35. Height of tulip plants grown for 56 days in a greenhouse medium
treated with sludge effluent or water (Kirkham 1977a).

Ammonium Toxicity. Prolonged application of ammonium-N may lead to physiological and morphological disorders of crop plants resulting in chlorosis and
necrosis of leaves and growth retardation (Maynard et al. 1966). Brown and
pitted necrotic leaf and stem lesions develop on tomatoes (Fig. 36) with the
severity directly related to ammonium concentration in the rooting medium.
Yields are reduced at high ammonium levels and free ammonium accumulates in
plants grown at high ammonium levels (Table 13).

Ammonium toxicity in tomato causes chloroplasts to become ellipsoidally
rounded and dispersed through the protoplasm rather than flattened around the
protoplasm (Puritch and Barker 1967). Vesicles develop in the plastids followed
by lamellae swelling and disappearance. Chlorophyll loss and photosynthetic
loss occur concurrently. Ammonium toxicity in cucumbers is indicated by injury
symptoms about one week after applying NH_4Cl (Matsumoto et al. 1968). Tissue
glucose content increases greatly while starch level decreases suggesting that
translocation of photosynthate is inhibited in ammonium toxicity. The leaf
symptoms include curling and marginal chlorosis.

Fig. 36.   Tomato stem sections showing no lesions (left) and severe lesion formation (middle), and leaflet with necrotic areas developing along midrib
and expanding to the lamina (right) (Maynard et al. 1966).

Table 13.   The influence of ammonium nutrition on growth and severity of
toxicity symptoms in 'Heinz 1350' tomato.

	Fresh wt.(g)	Stem length(cm)	Injury rating[y]	
			Stems	Leaves
.01 M $(NH_4)_2 SO_4$	61 ab[z]	33.5 ab	1.7 b	1.0 b
.05 M $(NH_4)_2 SO_4$	49 a	29.7 a	3.0 c	2.5 c
.01 M $NH_4 NO_3$	82 b	38.0 c	1.5 b	0.2 ab
.05 M $NH_4 NO_3$	71 ab	36.2 bc	3.0 c	2.2 c
.02 M $NaNO_3$	66 ab	36.3 bc	0 a	0 a
No N	59 ab	37.9 c	0 a	0 a

[y] Injury rating scale 0, no injury; 3, severe lesions.
[z] Mean separation within columns at the 5% level.

Data from Maynard et al. (1966)

CHAPTER 3

PESTICIDE POLLUTION

Herbicides, insecticides, fungicides, fumigants and other pest-control chemicals are a necessary part of horticultural crop production systems. As a group they are referred to as pesticides and members of each class have the potential to become pollutants and cause crop injury. Large amounts of herbicide are used on a world-wide basis and effective killing of weeds in crops depends on the differential sensitivity of weed and crop species to the herbicide as well as on the use of appropriate management systems designed to maximize herbicide injury to weeds with minimum or no injury to the crop. Insecticides are also used widely to control insects that reduce yield or quality of horticultural plants. Insecticide use has been severely restricted in view of the long-term undesirable ecological effects of some chemicals widely used at one time, including DDT, aldrin and dieldrin. Fungicides, used to protect horticultural plants from diseases caused by fungi, as a group have caused the least injury as air pollutants. Fumigants are volatile pesticides which kill a wide range of animals and plants. They can easily become air pollutants and must be used with great care. Other pesticides include rodenticides, nematicides and bactericides, which may also become pollutants when misused or allowed to move to non-target areas. The many effects of such pesticides on fruit and vegetable physiology have been compiled by the U.S. National Academy of Sciences (1968).

Pesticides can play a role as pollutants by directly injuring sensitive plants either as a result of drift at time of application or as residues which affect subsequent crops in the same area. Insecticides also behave as pollutants when they kill pollinating insects which, if this occurs at pollination time, may reduce yield. Pesticides, for example, arsenicals and dinitrophenols, and some synthetic organic insecticides, may also destroy insect parasites that control populations of economic pests (Anderson and Atkins 1968).

Dislodgable residues of pesticides on crops are of concern from the point of view of field employee exposure to pesticide residues. Application of several pesticides to citrus leaves and fruit followed by shaking the leaves with water and wetting agent at intervals up to 21 days showed that dislodgable residues decrease with time and/or rainfall (Thompson and Brooks 1976). There are marked differences among pesticides in amounts of dislodgable residues.

Many problems involving pesticide injury on horticultural crops are a result of the use of an inappropriate chemical for the particular species.

The pre-testing of pesticides on crops prior to approval is strongly recommended because there are deleterious effects of a number of combinations. For example, several pesticides reduce fruit set and yield of apples (Donoho 1964). Pesticides may be harmless to the horticultural species in production under some environmental conditions but harmful pollutants in a different environment. Recommendations thus often stipulate that application be made to a particular crop only if certain environmental conditions prevail. The injury symptoms often are a reflection of the release of a particular compound under certain conditions. For example, lime sulphur spray injury may have symptoms identica' to those produced by H_2S (McCallan et al. 1936).

3.1 Pesticides in Air and Pesticide Drift

The application of pesticides may be quite inefficient in many horticultural crops. A portion of the pesticide may fail to reach or remain on the target, and instead move in the air to other areas where it may injure other plants or affect animal life. This movement of pesticides as air pollutants is referred to as pesticide drift. There are many problems associated with drift during and after application of pesticides (Akesson and Yates 1964). The air pollution aspects of pesticide usage have been reviewed in a monograph format covering sources and transport of airborne pesticides, sampling and analysis methods, and assessment of toxicity with emphasis on toxicity to humans (Lee 1976).

Pesticides may become airborne as droplets, vapour or dusts during application, or later from soil, vegetation or water (Lunin 1971). Spray droplet size is a very important consideration with large droplets striking the target or falling to the ground. As size becomes smaller, the droplets are more readily diverted by air currents. Droplet diameters below 100 microns result in very slow fall rate or suspension in the air. Such small droplets can be readily carried to other areas. Low density dusts and volatile pesticides are also very susceptible to drift. Post-application entry into the air may take place as a result of adsorption on soil particles which enter the air as a result of wind action or mechanical agitation of dry soils as in cultivation. The pesticide may also volatilize from soil or plants, or adhere to particulates resulting from the burning of crop residues.

Reduction of pesticide air pollution requires a knowledge of the factors affecting drift. Spray droplets decrease in size after emission from the sprayer as a result of water evaporation and/or chemical volatilization. The rate of change depends on the original size of the droplets, the air humidity, the duration in the air, the velocity of the droplets, and other chemical and physical factors. As droplets decrease in size they become more subject to drift, moving upward with convection currents and horizontally with winds.

Pesticide drift at the time of application can be reduced by selection of equipment which applies sprays at lower heights using nozzles which provide suitable particle size with as few fine droplets as possible (Middleton 1965). In particular the sprayer nozzles should be matched to the size and shape of the target, and nozzles, pressures, and volumes used that will produce the best combination of coverage and reduction of fine driftable droplet sizes. Dusts should be avoided except those with suitable particle size. Timing of all applications should be restricted to weather conditions which limit the pesticide to the desired application site.

Pesticide behavior in the air is complicated by the simultaneous existence of pesticides in both particulate and vapour phases (Braidenbach 1965). As far as possible, pesticide formulations which reduce volatility should be used, especially for herbicides that can injure sensitive crops downwind. For example, 2,4-D amine would be preferable to 2,4-D ester because the amine is less volatile.

Pesticide drift can be substantial. For example, the determination of parathion content of air in and near a Washington State, U.S.A. orchard showed that the concentration varied from $0 - 5.53$ mg/m^3 near the loading and mixing area, from $0 - 0.02$ mg/m^3 at a residence near the treated orchard, and from 0 to trace amounts at a residence some distance from the sprayed orchard (Middleton 1965). Drift of pesticides applied by airplane has also been studied. Insecticides can be detected under adverse conditions as far as 800 m from the line of application (Argauer et al. 1968).

The danger of herbicide drift injury has resulted in the establishment of regulations in some countries restricting the use of some phenoxy compounds. The regulations stipulate the permitted time of use and method of application.

3.2 Pesticides in Soil

A survey of orchard and vineyard soils in Ontario, Canada has revealed the extent of accumulation and persistence of many inorganic pesticides and some organochlorine and organophosphorus pesticides (Frank et al. 1976b). Arsenic, Pb, Hg and Cu are elevated as a result of years of use of these elements as fungicides (Table 14). Organochlorine residues, such as the no longer used DDT, are substantial but only traces of some organophosphorus pesticides are found.

Pesticide behavior in soil can be simulated in a computer model to study the rate of leaching through the upper meter of soil after spring application to a potato crop growing in a sandy loam (Leistra and Dekkers 1976). Calculated leaching is dependent on decomposition rates and adsorption strengths, and

Table 14. Pesticides containing metals recommended in Ontario, Canada (Ontario
Ministry of Agriculture and Food recommendations for 1892 - 1975).

Chemical	Metal composition of product	Period of recommendation	Crops
Insecticides			
Copper aceto-arsenite	2.3% As, 39% Cu	1895 - 1920	Apples and cherries
(Paris green)		1895 - 1957	Vegetables and small fruits (foliar and bait)
Calcium arsenate	0.8 - 26% As	1910 - 1953	Fruit and vegetables
Lead arsenate	4.2 - 9.1% As	1910 - 1975	Apples
	11 - 26% Pb	1910 - 1971	Cherries
		1910 - 1956	Peaches
		1910 - 1955	Vegetables
Mercuric chloride	6% Hg	1932 - 1954	Cruciferous crops
Zinc sulfate	20% Zn	1939 - 1955	Peaches
Fungicides			
Copper sulfate -			
calcium salts	4 - 6% Cu	1892 - 1975	Fruit and vegetables
(Bordeaux and Burgundy mixtures)			
Fixed copper salts	2 - 56% Cu	1940 - 1975	Fruit and vegetables
Ferbam	0.5 - 12% Fe	1947 - 1975	Vegetables
		1948 - 1975	Fruit
Maneb	1 - 17% Mn	1947 - 1975	Fruit and vegetables
Mancozeb	16% Mn, 2% Zn	1966 - 1975	Fruit and vegetables
Methyl and phenyl			
mercuric salts	0.6 - 6% Hg	1932 - 1972	Seed treatment
Phenyl mercuric acetate	6% Hg	1954 - 1973	Apples
Zineb and ziram	1 - 18% Zn	1947 - 1975	Vegetables
		1957 - 1975	Fruit
Topkiller			
Calcium arsenite	30% As	1930 - 1972	Vegetables
Sodium arsenite	26% As	1930 - 1972	Vegetables

Data from Frank et al. (1976a)

ranges from virtually none to 10% of the dosage or more for compounds with high
persistence and mobility. Both decomposition and uptake by plants are important
factors in reducing leaching.

Application of PMA sprays to an apple orchard followed by soil residue
examination shows that Hg residues are almost entirely confined to the top five
cm of soil (Ross and Stewart 1962). Dieldrin lost from the soil is found in
leaves which absorb the pesticide from the vapour phase (Caro and Taylor 1971).
The nitroaniline herbicide benefin readily volatilizes and undergoes photo-

decomposition (Eshel and Katan 1972c).

3.3 Pesticides in Water

Pesticide content of water will be of great importance in determining the use of that water. Factors involved are the relative occurrences and levels of pesticides in solution, adsorbed on solids, or present in the flora and fauna (Breidenbach 1965).

An example of the importance of water quality is a reported occurrence of severe growth regulator injury to glasshouse tomatoes (Williams et al. 1977). The injury is characterized by failure of leaf expansion, inrolling of leaf margins, and changes in leaf venation and margin morphology, along with fruit deformation. The injury could be traced to the domestic water supply which had been contaminated by effluent from a factory manufacturing 2,3,6-TBA.

3.4 Persistence of Pesticides

In general, the amount of pesticide residues in plants can be controlled by timing of sprays (Edwards 1970). Most of a substantial residue in plant tissue at time of harvest is derived from foliar sprays applied near harvest time. This relationship makes it easy to control and limit residues of pesticides in food crops.

Numerous studies of pesticide persistence have been conducted. For example, residues of disulfoton and phorate, systemic organophosphorus insecticides, can be detected in pea vines but not in shelled peas (Chisholm and Specht 1967). The insecticides do not affect yields or maturation rate, or produce injury symptoms.

The fungicides which persist longest are those that contain heavy metal trace elements. Many of these break down to residues containing Cu and Hg which may be harmful to soil organisms. For example, application of large amounts of copper sulphate over many years to orchard soils may result in severe reduction in numbers of earthworms together with very poor structure in orchard soils. If other crops are planted on these soils Cu toxicity chlorosis and high tissue Cu content may occur.

The organic Hg fungicides are very important as persistent pollutants. The use of alkyl-Hg compounds as seed treatments from about 1940 onward resulted in increased Hg accumulations in birds. Prohibition of alkyl-Hg compounds about 1966 and their replacement by alkoxyalkyl-Hg compounds as seed dressings resulted in an abrupt decline to a very low level of Hg in predatory bird species (Johnels and Westermark 1969). However, alkoxyalkyl-Hg can be converted to methyl-Hg, a very mobile form of this metal.

Herbicides are generally much less persistent than the chlorinated hydrocarbon insecticides (Kearney et al. 1969). Most herbicides break down rapidly
in the soil but a number may persist for a year or longer including atrazine,
dichlorobenil, diuron, fenuron, monuron, neburon, picloram, propazine, propham,
simazine, trichloroacetic acid, and 2,3,6-trichlorobenzoic acid. Injury to subsequent crops may be possible with the use of such herbicides and the subsequent
plantings should be of tolerant species. Such herbicide pollution is a particular problem when land is replanted quickly after a crop failure due to frost,
hail or other destructive agents.

3.5 Herbicides

Inadequate herbicide tolerance may result in occasional injury to a crop.
One explanation is that environmental factors affect crop response. This is
shown in experiments on diphenamid uptake from soil by eggplant, pepper and
tomato (Eshel and Katan 1972b). In all species an increase in temperature during
the light period from 20 to 30°C increases phytotoxicity while a temperature
increase from 15 to 20°C decreases diphenamid toxicity. Tolerance is thus markedly affected by temperature with reduced tolerance at high day and low night
temperatures.

Timing of herbicide application is an important consideration in minimizing
injury to the crop plants. For example, delaying application of diphenamid from
the day of sowing to close to emergence reduces toxicity to the crop while only
partially reducing herbicidal action (Eshel and Katan 1972a). The delay allows
the crop's root system to penetrate below the soil layer containing the diphenamid. An increase in Rhizoctonia damping-off due to diphenamid is also reduced
by delayed applications. The interaction with Rhizoctonia is attributed to
suppression by the herbicide of antagonists to the fungus (Eshel and Katan
1972c). Increased incidence of damping-off diseases caused by various soil
pathogens can be noted after diphenamid is used as a pre-emergence herbicide
for peppers and tomatoes.

Studies of herbicide sensitivity of horticultural crops are conducted to
gain information on potential problems. An example is the application of 15
herbicides to the roots of six deciduous fruit species growing in sand (Lange
and Crane 1967). Of pre-emergence herbicides, trifluralin, DCPA, diphenamid and
EPTC are most promising, while among post-emergence herbicides, dalapon, amitrole
and paraquat are the most promising for use in orchard crops from the point of
view of least injury to the fruit trees. 2,4-D at low rates is the safest growth
regulator type herbicide for orchard use of those tested.

Peach is more tolerant than apricot to two triazine herbicides, simazine and prometryne (Tweedy and Ries 1966). The rootstock does not alter the tolerance and the tolerance of peach is attributed to detoxification in the scion.

Phenoxy Herbicides. An outstanding example of a pesticide becoming a pollutant is the volatilization and drift of certain phenoxy herbicides, including 2,4-D and 2,4,5-T. The adverse effects of such compounds are observed on sensitive crops such as grape and tomato at distances as much as 25 km from the site of application. The physical state, particle size, and extent of area treated are important considerations in the use of such herbicides, as well as the volatility of the chemical. Compounds of high volatility may injure sensitive species (at considerable distances) for several weeks.

Studies of 2,4-D as an air pollutant on horticultural crops reveal the wide range of sensitivity and symptoms (Weigle et al. 1970). Numerous injury symptoms are attributed to 2,4-D including deformed fruit on strawberries, uneven berry ripening in grapes, blotchy ripening and deformed fruit in tomatoes, and decreases in yield of grapes and strawberries. 2,4-D may enhance fruit set of tomato but the cumulative effect of repeated exposure is severe plant injury, reduction in productivity, and increase in blossom-end rot. In commercial practice, tomatoes are affected by volatile drift of 2,4-D near areas of its use in weed control. Plants may appear to recover from a single exposure and yield satisfactorily so damage is difficult to assess.

2,4-D effects on horticultural plants can be studied outdoors using concentric rings of plants around a 2,4-D source (Sherwood et al. 1970). 2,4-D causes such diverse responses as increased overwintering injury, carryover effects on potted plants moved elsewhere, and decreased effects on virus-infected plants. Among symptoms are increased stiffness of leaves and soft shoots, fewer and smaller flowers, shorter and thicker fruits, curling of bean pods, and unusually waxy woody plants (Figs. 37,38,39).

2,4-D at low concentrations may decrease fruit drop from apricot trees but a residue is left in the fruit. Analysis of apricot fruit reveals that not much 2,4-D is lost over a five to seven week period but the amount found is well below usual residue limits (Love and Donelly 1976).

Widespread injury due to 2,4-D and 2,4,5-T may be observed in an urban-industrial area. For example, such injury was traced to a local manufacturer of these herbicides (Lanphear and Soule 1970). Chemicals involved in the manufacture as well as the two herbicides are injurious to plants. A number of urban plants are sensitive, according to tests of these chemicals performed initially on forsythia, tree-of-heaven and redbud plants (Tables 15,16,17). In some cases the only symptom is curling of the leaf margins, but in more severe cases there is stem epinasty, defoliation, and necrosis of meristematic tissue.

Fig. 37. The diffusion tests circle used for 2,4-D exposures (Sherwood et al. 1970).

Table 15. 2,4-D and 2,4,5-T sensitivity of urban plants.

Very sensitive	– tree-of-heaven
	– Eastern redbud
	– Chinese elm
Sensitive	– platane maple
	– silver maple
	– red ash
	– showy border forsythia
	– ginkgo
	– tulip tree
	– tatarian honeysuckle
	– old fashioned weigela
Insensitive	– sugar maple
	– barberry
	– apple
	– cherry and plum
	– common lilac

Data from Lanphear and Soule (1970).

Fig. 38. Severe epinasty of grape at 9 m from the 2,4-D source (Sherwood et al.
1970).

Table 16. Initial screening of suspected industrial chemicals on forsythia at
four concentrations.

	Injury index [z]			
	Concentration (ppm)			
Chemical	1000	100	10	1
Sodium salt 2,4-D	4	4	1	0
2,4-D isobutyl	4	4	4	2
2,4-D acid	4	4	2	1
2,4-D isoctyl	4	4	1	1
2,4,5-T isobutyl	4	4	1	1
2,4,5-T isoctyl	4	2	1	1
2,4,5-T acid	4	4	1	0
2,4-dichlorophenol	0	0	0	0
Parachlorophenol	0	0	0	0
Orthochlorophenol	0	0	0	0
Phenol	0	0	0	0
Monochloroaceticacid	0	0	0	0
Pentachlorophenol	0	0	0	0
Sodium pentachlorophenol	0	0	0	0

[z] Injury rating system: no injury = 0; slight leaf curling = 1; moderate leaf
curling = 2; severe leaf curling = 3; severe leaf curling, petiole and stem
distortion = 4. Data from Lanphear and Soule (1970).

Fig. 39. Redbud leaves and shoots at 9 m from the 2,4-D source (Sherwood et al.
1970).

Table 17. Secondary screening of tree-of-heaven and redbud with chemicals
injurious to forsythia.

	Injury index [z]								
	Tree-of-heaven					Redbud			
	Concn (ppm)					Concn (ppm)			
Chemical	100	10	1	0.1	0.01	10	1	0.1	0.01
Sodium salt 2,4-D	4	2	0	–	–	1	–	–	–
2,4-D isobutyl	4	2	1	0	0	1	1	0	0
2,4-D acid	4	2	1	–	–	3	1	–	–
2,4-D isoctyl	4	2	0	–	–	0	0	–	–
2,4,5-T isobutyl	1	0	0	–	–	0	0	–	–
2,4,5-T isoctyl	1	0	0	–	–	1	0	–	–
2,4,5-T acid	2	1	0	–	–	1	0	–	–

[z] For injury rating system see Table 16.

Data from Lanphear and Soule (1970).

3.6 Pesticide-induced Abnormalities

Many anatomical, cytological and genetic abnormalities can be induced by pesticides. For example, spraying immature onion inflorescences with dithane fungicides results in production of abnormalities in pollen mother cells with different formulations having different intensities of effect, apparently due to the Zn, Na and Mn contained in different formulations (Mann 1977). Erratic behaviour of chromosomes at meiosis causes pollen sterility and reduces seed set. An example of a different abnormality is that EPTC inhibits epicuticular wax production on developing leaves of cabbage (Flore and Bukovac 1976).

3.7 Metabolism of Pesticides

Studies of the rate of metabolism of di-syston, a systemic insecticide, show that metabolism and hydrolytic decomposition of the toxic products occur two or three times as fast in tomato as in cotton (Metcalf et al. 1959). Rates of oxidation of metabolites increase about 1.9 times for each $10^{o}C$ rise in temperature.

Herbicides like simazine combined with amitrole-T may influence N metabolism of apple or peach trees (Ries et al. 1963), stimulating growth and increasing plant N content compared with supplemental N treatments, no weed control, hand hoeing, or use of black plastic mulch. A direct influence of the combined herbicides on N metabolism of the trees is thus indicated. Simazine alone incorporated in soil at 2.5 ppm reduces shoot and root growth of apple (Dvorak 1968).

CHAPTER 4

WATER POLLUTION

The subject of water pollution in general is vast and complex and beyond the scope of this presentation. Only the direct relevance of water pollution to horticulture will be introduced here. The utilization of water for irrigation of horticultural crops may be limited or impossible because of impurities, particularly ions of salts (Wehrmann 1974) and other chemical contaminants (Barker and Craker 1973). Polluted water is usually of much greater significance in glasshouse production since glasshouses depend on irrigation exclusively compared to the outdoor situation where irrigation normally is used only to supplement natural precipitation. In glasshouse production large amounts of irrigation water are used per unit area, because of shallow soils in containers and raised benches and because of high leaching requirements when nutrient solutions are added. A consequence of the use of large amounts of water in glasshouse production may be the contamination of nearby surface and ground water supplies by leachate and nutrient solutions. Additional concerns in glasshouse production are the quality of water for mist propagation, for use in evaporative coolers, and for conditioning cut flowers before shipment. Carbonates and bicarbonates can result in foliar deposits and clogging of mist nozzle orifices in propagation. Evaporative fan and pad cooler systems will concentrate salts which will ultimately clog the pads. Water quality for conditioning flowers is very important. Even small amounts of dissolved salts will adversely affect the keeping quality of flowers. Deionizing equipment may be needed for improving the quality of conditioning water. Species tolerance to salinity varies widely and, for example, among glasshouse species, chrysanthemums are quite tolerant while some azalea cultivars are very sensitive.

There are suggested critical values for water salinity above which yield depressions would be expected to occur. Species are grouped into four classes of relative tolerance (Table 18).

There are a number of criteria of water quality for horticultural purposes and the need for each can be justified (Wehrmann 1974). Total salt determines water uptake by the plant. Sodium and Ca affect soil structure and pH. Waters with high $CaCO_3$ or high Fe content can result in salt precipitation on leaves. Boron toxicity may occur if there is accumulation from water. Species differences are also significant. For example, cucumber is more sensitive to Cl^- than tomato and more sensitive in a sandy than in a clay soil. Leaching with

Table 18. Suggested critical values of irrigation water salinity for protected cultivation.

| | | Relative tolerance of plants to salt | | | |
| | | very low | low | medium | high |
		Orchid Fern	Araceae Gesneriaceae Cucumber	Begonia Cyclamen Lettuce	Carnation Tomato Cabbage
Salt	mg/l	250	500	750	1000
EC	mmhos/cm	0.4	0.8	1.0	1.5
Cl^-	mg/l	50	100	200	300
$SO_4^=$	mg/l	100	200	250	300
Na^+	mg/l	50	100	150	150
Hardness	me/l				
– total		2	4	7	11
– temporary		1	3	5	9
Boron	mg/l	0.5	0.5	0.5	0.5
Iron	mg/l	1	1	1	1

Data from Wehrmann (1974)

excessive irrigation water diminishes the differences between soils. The useful-ness of river water for greenhouse irrigation is dependent on salt content with densely populated areas resulting in river water with high total salt and NaCl contents. There may be NaCl introduction to river water as a waste product of K fertilizer production as well as seasonal fluctuations in salt content. Other water pollution problems that may affect horticultural crop production include excessive trace element or acid content, and the presence of disease organisms or insects.

CHAPTER 5

SALT POLLUTION

Excessive accumulation of soluble salts is an important factor in irrigation of horticultural crops in many parts of the world. Additional salt problems
may arise in sea coast areas where salt spray injury and salt water intrusions
may occur. Salt injury may also occur from the use of road salt in winter in
some areas for ice control. Another horticultural problem is the rapid accumulation of soil salts in glasshouse operations.

Salt spray or salt aerosol research is of particular interest to those
involved in coastal horticulture or to home owners near the seashore. An additional dimension has resulted from the use of cooling towers to dissipate heat
from electric generation plants in coastal areas. These may utilize saline water
and the drift from these cooling towers may contain salt aerosols similar to sea
salt spray.

Sprinkler irrigation with water containing salts may result in accumulation
of salt residues on foliage followed by injury when the residues redissolve in
spray and dew. The effect may be injury similar to that occurring along sea-
coasts or roadways where salt-spray accumulations can occur. Foliar absorption
of ions followed by injury may occur in species sensitive to one or more of the
ions, even without salt accumulation on the leaf surface.

Soil salinity increases as a result of fertilizer additions and use of salt-
containing irrigation water. Together they determine the salt content of the
soil solution which in turn determines the potential for salt injury. For most
plants the major undesirable effect of soil salinity is the reduction in water
availability resulting from increased osmotic concentration of the soil solution.

Defining salt tolerance is difficult and requires consideration of growth in
terms of annual or perennial habit. Perennials are susceptible to cumulative
effects over a period of years. Growth rate and survival may be less important
than salt effects on economic yield. Plant-to-plant variability is also a factor
in making it difficult to assess salt responses objectively.

One approach to overcoming salinity problems is to understand the physiology
of salt tolerance in plants. This would include evaluation of the physiological
factors in terms of osmotic effects and specific ion effects, both nutritional
and toxic (Bernstein and Hayward 1958). The effect of salinity may differ at
different stages of development. Salinity exerts effects on development and

yield depending on the type of crop, the growth habit, and the stage of development. Classifications have been developed for salt tolerance of various horticultural species grouped as vegetable crops, fruit crops, ornamentals and flowers, and shade and roadside trees (Hayward and Bernstein 1958).

The development of salt insensitive crops for use in salt-prone regions is an objective of some research programs. For example, there are sources of salt insensitivity in tomato species (Rush and Epstein 1976). A relatively salt-insensitive species from the Galapagos Islands has low levels of K and Na, total amino N, specific amino acids, and free amino acids, compared to a sensitive commercial cultivar.

5.1 Salt Injury to Vegetable Crops

Salt accumulation may be a serious and costly problem in vegetable production in countries in which salt-laden water is utilized in irrigation. Salt bands may accumulate near the surface of beds of row crops and seeds may fail to germinate. Variation in salt concentrations may also cause variability in maturity. A recommended approach is to sprinkle irrigate with high quality water to ensure salt removal from surface layers, and use precision planting which results in earlier maturity, better quality and higher yields (Robinson and McCoy 1965). The range of vegetable crop sensitivity to salt is from beets, kale, spinach and asparagus which have low sensitivity, to beans, celery and radish which have high sensitivity (Hayward and Bernstein 1958). The approximate order of increasing salt sensitivity by vegetable crops of intermediate sensitivity is tomato, cruciferous crops, sweet corn, lettuce, potato and onion.

Bush beans have been exposed to saline mist to check relative humidity and particle size effects (McCune et al. 1977). Increasing either results in increased toxicity of the mist.

5.2 Salt Injury to Fruit Crops

Fruit crops are generally much more sensitive to Na^+ and Cl^- than vegetable crops. Salt sensitivity ratings for fruit crops have been well established by many investigators but interacting climatic factors may affect the relative position of species. The date palm, mulberry, olive, pomegranate and jujube appear to have the lowest salt sensitivity while the most sensitive species include papaya, avocado, loquat, persimmon, blackberry and gooseberry (Hayward and Bernstein 1958). Pear, apple and most stone fruits are considered to have high salt sensitivity while citrus are moderately sensitive, with lemon more sensitive than orange and grapefruit. Among temperate fruit tree crops, peaches are the most susceptible to salt injury. Weakened portions of salt-injured trees may be attacked by peach canker disease.

Irrigation waters of graded salinity have been applied to fruit trees to determine rates of foliar accumulation and leaf injury (Ehlig and Bernstein 1959). Foliar absorption of Na^+ and Cl^- is most rapid in apricot and almond, intermediate in plum and orange, and very slow in avocado. Plum leaves are the most sensitive to these ions. Almond, apricot and plum leaves develop tip-burn and orange leaves lose their gloss and colour and sometimes develop tip-burn. In contrast, vegetable crops are not injured by sprinkling with similar salinized irrigation water. Both exposure to salt spray near roadways on which deicing salt is used or experimental sprays cause similar flowering suppression and tip dieback in apple (Hofstra and Hall 1971).

5.3 Salt Injury to Ornamentals and Shade and Roadside Trees

The nature of road salt spray injury to trees and other plants near highways has been described (Westing 1969; Hofstra and Hall 1971). Injury and tissue Na and Cl contents are greatest on the downwind side of the highway, on the windward side of the plants, and on plants in exposed positions. Deicing salt is apparently whipped up in a spray by traffic and blown onto vegetation. Other considerations are the extent of injury, salt behavior in soil, uptake, translocation and accumulation, effects on the plants, and amelioration methods. It has been suggested that Cl^- or Na^+ accumulation in leaves of shrubs and other landscape plants may cause injury by interfering with normal stomatal closure, causing excessive water loss and drought-like leaf injury symptoms (Bernstein et al. 1972).

Salt spray injury symptoms have been observed on 75 species near roadways and exposed to aerial drift of deicing salt (Lumis et al. 1973). Susceptible deciduous species have twig dieback and inhibition of flowering (Fig. 40). Conifers have needle browning, premature needle abscission and twig dieback. Insensitivity is associated with resinous buds, submerged buds, and needle cuticular wax.

Sodium and Cl contents in twigs of roadside trees and shrubs exposed to aerial drift of road salt vary from winter to spring (Lumis et al. 1976). Sodium content is somewhat lower than Cl. Both increase until March and then decline. Levels are higher in buds than in adjacent twig tissues. Salt content is related to plant injury, to relative tolerance of the species, and to temperature and snowfall during the study period. Analysis of both soil and hemlock needles near a highway through one season shows the effects of use of deicing salt (Langille 1976). Soil Na increases up to 12 m from the highway and soil Cl up to 61 m. Tissue Na and Cl increase up to 61 m from the highway.

The deleterious effects of various salts differ according to studies on

Fig. 40. 'Red Osier' dogwood branch with new growth arising only from 2 year old wood as a result of salt spray injury (Lumis et al. 1973).

English ivy (Dirr 1975). Shoot dry weights are decreased and leaf margin chlorosis is caused by application of Cl salts, but SO_4 salts only decrease shoot weights without causing necrotic symptoms (Table 19). Injury is more severe

Table 19. Effect of salts and application method on growth of English ivy.

| | Dry weight (gm/pot) | | | |
| | Shoots | | Roots | |
	Soil	Spray	Soil	Spray
Control	11.08 c[z]	11.45 c	2.72 ab	3.01 ab
Na_2SO_4	6.08 b	11.60 c	2.56 ab	3.54 b
K_2SO_4	5.06 ab	10.82 c	2.46 ab	2.74 ab
NaCl	2.96 a	3.30 ab	2.14 a	2.26 a
$CaCl_2$	4.78 ab	3.07 ab	2.43 ab	2.16 a
KCl	3.33 ab	3.60 ab	2.32 ab	4.43 ab

[z] Mean separation within shoots or roots by Duncan's multiple range test, 5% level.

Data from Dirr (1975).

when Cl salts are sprayed, compared to application as a soil drench. Root
weights are generally not reduced by the salt application. This study included
sufficient factors to illustrate that evaluations of salt-induced injury should
include consideration of the salt used, concentrations, application methods,
osmotic effects and shoot tissue Cl.

The decline of maple trees in temperate regions is a puzzling phenomenon
and has resulted in a number of salt-oriented studies. Maple decline in New
Hampshire, U.S.A. is attributed to road salt because the disorder is confined
to roadside areas and tissue Na levels are elevated (Lacasse and Rich 1964).
The principal toxic ion inducing injury to sugar maple is considered by some to
be Cl^- (Holmes and Baker 1966). Necrosis or death occurs when a tissue Cl concentration of 0.5 to 1.5% is reached. However, others find that leaf injury on
sugar maples is correlated with both tissue Na and Cl (Hall et al. 1973) with
no evidence that deficiencies of essential elements are induced. Endomycorrhizae
of sugar maples are affected by salt according to a study of deicing salt
effects on roadside tree dieback (Guttay 1976). Progressive dieback is correlated
with decrease in endomycorrhizae, increased average depth of viable roots, and
increased Na and Cl in the soil.

After seven years of application of NaCl and $CaCl_2$, it has been concluded
that winter road salting is not very harmful to trees in Massachusetts, U.S.A.
(Holmes 1961). Some maples, a birch and a pine died after many salt applications.
White and black oak are not affected by salt applications and do not have elevated tissue Cl^- in contrast to other species. Run-off water high in NaCl or
$CaCl_2$ does not cause injury to sugar maples although leaves contain three to
six times as much Cl^- as control trees.

Relationships of salt spray to woody plant winter hardiness have been
established. Highway and garden grown lilac and green ash have different cold
hardiness patterns (Sucoff et al. 1976). Hardiness is the same for both sources
of plants in November but from mid-December to March, highway twigs are less
winter hardy. The differences coincide with twig Cl content attributable to
salt spray drift onto the plants growing near a highway. Decreased winter
hardiness of 'Radiant' crabapple and lilac is a result of NaCl application
whether plants receive NaCl from controlled applications or from highway deicing
salts (Sucoff and Hong 1976). In either type of application the loss of hardiness is proportional to the increased tissue Cl and Na concentrations (Table 20).

Suggestions for amelioration of road salt injury include use of Ca salts
rather than Na salts, even though Ca salts are more expensive; removal of sensitive species from near roadways; and the use of more non-sensitive species such
as grasses in the most vulnerable areas (Westing 1969).

98

Table 20. Highest killing temperature (HKT $^{\circ}$C) and stem Cl (% dry weight) for salt- and water-treated 'Radiant' flowering crabapple collected from highway (H) and garden (G) locations.

Twig position	Variable	Treatment	Nov. 11		Dec. 10		Jan. 7	
			G	H	G	H	G	H
Tip	HKT	Water	−21	−30	−30	−30	−45	−30[z]
		Salt	−21	−30	−21	−21	−21	
	Cl	Water	0.00	0.00	0.00	0.15	0.00	0.67
		Salt	0.00	0.00	0.40	0.73	0.64	1.18
Base	HKT	Water	−30	−30	−45	−45	−45	−30
		Salt	−30	−30	−30	−21	−45	−21
	Cl	Water	0.00	0.00	0.00	0.09	0.00	0.45
		Salt	0.00	0.00	0.22	0.35	0.38	0.67

[z] 97% stem injury at 4°C.

Data from Sucoff and Hong (1976). Reproduced by permission of the National Research Council of Canada from the Canadian Journal of Botany, 54, 2816-9.

CHAPTER 6

CELLULAR AND BIOCHEMICAL RESPONSES TO POLLUTANTS

The modes of action of pollutants in causing abnormal plant responses are not well understood. The uptake of air pollutants is considered to take place largely in the leaf tissues of the plant, primarily by diffusion through stomata. Soil and water pollutants generally enter through the roots although leaf absorption of all kinds of pollutants does occur. Most pollutants probably act at the cellular and metabolic level, affecting structural components and metabolic processes. A number of studies of ultrastructural responses have provided some indications of sites of action. Plant membranes are physically affected in many cases, but the biochemical basis for the membrane disruptions has not had much definitive work. An understanding of biochemical modes of action will be helpful in research on the chemical and genetic bases for pollutant resistance or protection. Most metabolic studies have been concerned with leaf tissue responses and little attention has been given to stem, flower, fruit or root tissues, probably because the leaves are assumed to be the dominant pollutant absorbing and metabolizing organs.

Reviews of the chemical, biochemical and physiological modes of toxic action of common air pollutants indicate that it is not yet possible to describe with certainty the detailed sequence of events leading to injury symptoms, in spite of a large number of research reports on mode of action studies (Mudd 1972). Some investigators have attempted to describe the mode of action of particular pollutants. For example, the sequence of O_3 injury is viewed as sulfhydryl oxidation, lipid hydrolysis, cellular leakage, lipid peroxidation, and cellular collapse (Rich and Tomlinson 1974). Other approaches include consideration of the effects of air pollutants on the biosynthesis of pharmaceutically valuable compounds (Jonas 1969). The implications of known effects on metabolism can be applied to synthesis of medicinal compounds.

Horticultural plants have been used frequently to establish the principles involved in cellular responses to pollutants. The present survey attempts to present the broad aspects of cellular and biochemical responses of all types of plants and to highlight some of the studies which involved horticultural plants. Information is assigned to sections but there is much overlap such as, for example, in the separation of enzyme effects from non-photosynthetic metabolism.

6.1 Cell Permeability

Oxidants. Considerable effects of O_3 on cell permeability have been reported (Evans and Ting 1973; Perchorowicz and Ting 1974). Following O_3 exposure the membrane permeability of bean leaves to tritiated water decreases, but to internal solutes and labeled rubidium and glucose permeability increases. Thus cellular membranes might be a primary target for O_3. Others also noted that O_3 increases cell permeability and causes changes in membrane components (Tomlinson and Rich 1971). Potassium plays an important role in maintenance of plant water status and O_3 interferes with the normal K ionic balance (Heath et al. 1974). The O_3 effect on SH groups and fatty acids may result in permeability changes with irreversible loss of both K^+ and water. There are thus many aspects to O_3 effects on cell membrane permeability (Ting et al. 1974). It is generally concluded that cell membranes are the initial targets for O_3 injury and that subsequent symptoms, including dessication, are a result of early changes in membrane properties.

Evidence for a repair mechanism for O_3-induced injury to membranes in 'Pinto' bean leaves has been obtained (Sutton and Ting 1977). Uptake of a-deoxyglucose by leaves is enhanced by O_3 exposure. Return to control levels takes five days in the light but is delayed in the dark.

Sulphur dioxide. An effect of sulphite on membrane permeability has been reported (Luttge et al. 1972; Puth and Luttge 1973). Sulphite can "seal" membrane leaks caused by n-butanol applied to beet roots. Membrane permeability is determined by the efflux of B-cyanin.

6.2 Metabolic Effects

A broad review of air pollutant injury to plant growth and development processes, photosynthesis and respiration, enzymes and coenzymes, and cellular components is available (Taylor 1975), as is a book devoted largely to O_3 and its effects on plant processes and constituents, with special attention to cellular membranes and biochemical reactions (Dugger 1974). Metabolic system effects of air pollutants have also been reviewed (Dugger and Ting 1970; Verkroost 1974; Ziegler 1973b). A number of physiological bases for oxidant response have been reported, including permeability changes prior to appearance of visible symptoms, changes in leaf carbohydrates, protective effects of red light against PAN injury to bean plants, correlation of sulfhydryl (SH) content in bean leaves with age and light regime, and age and light effects on metabolism of labeled PAN by bean leaf tissue (Dugger et al. 1966). An extensive listing of the most important known metabolic effects of some of the common air pollutants has been prepared with particular attention to changes in levels of enzymatic activity after pollutant exposure (Horsman and Wellburn 1976). The effects of O_3 on the energetics of plant systems have been reviewed by Pell (1974). Differences in stage of symptom development and nature of O_3 exposure are factors resulting in variable responses of plant system energetics. In general, O_3 causes decreases in oxidative- and photo-phosphorylation and increases in ATP and total adenylate content of plant tissue.

Enzyme Activity. The role of enzyme studies in determining the mode of action of air pollutants can be discussed as a result of studies with pea and tomato tissue (Wellburn et al. 1976). Some enzyme systems are stimulated by some pollutants and inhibited by others. Some pollutant mixtures are also tested and some interactions result in additive or synergistic effects.

Oxidants. Activity of many peroxidase isoenzymes in bean leaves is increased by O_3 (Curtis and Howell 1971), even in leaves which have no visible injury symptoms. There is increased peroxidase and cellulase activity in O_3-exposed bean leaves (Dass and Weaver 1972). Ozone exposure of soybean leaves initially depresses nitrate reductase activity as well as reducing sugar and amino acid content (Tingey et al. 1973c). The O_3 depression of nitrate reduction is due to interference with the supply of NAD(P)H for the reaction. Nitrite reductase activity and chlorophyll content of corn seedlings are sensitive to O_3 (Leffler and Cherry 1974). A two-phase response to O_3 is hypothesized, with the first phase at the chloroplast level and sensitive to low O_3 levels. The second phase is considered to be at the cellular level and more O_3-tolerant.

PAN inactivates isocitric dehydrogenase, G-6-P dehydrogenase and malic dehydrogenase (Mudd 1963). PAN oxidizes sulfhydryl groups, inactivating these enzymes but not ribonuclease which has no sulfhydryl groups.

Sulphur dioxide. Sulphite is an effective inhibitor of ribulose-1, 5-diphosphate carboxylase in spinach chloroplasts (Ziegler 1972). Phosphoenol pyruvate carboxylase in corn leaves (Ziegler 1973a) and spinach leaves (Mukerji and Yang 1974) can also be inhibited by sulphite. Sulphite inhibits various forms of malate dehydrogenase in spinach and corn (Ziegler 1974a,b). Many oxidizing enzymes, including phenolase, ascorbate oxidase and pyruvate decarboxylase, are inactivated by SO_2 (Haisman 1974). In SO_2-treated peas the mitochondrial form, but not the cytoplasmic form, of glutamate oxalacetate transaminase is inhibited (Pahlich 1973). Glutamate dehydrogenase is activated in SO_2-treated peas (Pahlich 1975). Exposure of pea seedlings to SO_2 and/or NO_2 reveals effects on enzyme activity (Horsman and Wellburn 1975). A synergistic effect is found for change in enzyme activity of ribulose diphosphate carboxylase and peroxidase within certain concentration ranges of the two pollutant gases.

Fluoride. Peroxidase and cytochrome oxidase activity in HF-exposed geranium leaves can be followed histochemically (Poovaiah and Wiebe 1971). Highest activity is near the injured parts of leaves, particularly in the phloem region.

6.3 Photosynthesis

Photosynthesis is suppressed by many air pollutants, including O_3, NO_2, NO, SO_2, HF and Cl_2 (Bennett and Hill 1974). Photosynthesis is also suppressed by relatively low concentrations of some pollutant combinations.

Oxidants. Species differences in ozonated hexene response were noted by Todd (1958). Concentrations which reduced photosynthesis and increased respiration in bean leaves had no effect on these processes in citrus leaves. The effects in bean were directly associated with visible leaf injury but respiration rate increased in Valencia orange leaves without visible injury. Plant growth suppression resulting from O_3 exposure is associated with partial stomatal closure and reduction in apparent photosynthesis and transpiration (Table 21). The reduction in apparent photosynthesis is often reversible with little or no injury development (Table 22). Ozone exposure reduces the total chlorophyll content of pea chloroplasts and is particularly effective in reducing chlorophyll b (Nobel 1974).

Table 21. Stomatal closure and reduction in apparent photosynthesis caused by 60 pphm O_3 for 1 hr.

| | Ave stomatal width microns | | % reduction in apparent photosynthesis |
	Control	O_3	
Bean	2.1	1.1	29
Potato	3.9	1.5	35
Tomato	2.8	0.6	25

Data from Hill and Littlefield (1969). Reprinted with permission from Environ. Sci. Technol., 3, 52-6. Copyright by the American Chemical Society.

Inhibition of photophosphorylation in pea chloroplasts is an effect of O_3 exposure (Nobel and Wang 1973). This is related to an increase in membrane permeability of the chloroplasts as indicated by decreases in the reflection coefficient of chloroplast membranes. Apparatus necessary for exposing chloroplast suspensions to air pollutants has been described in detail (Coulson and Heath 1974). Using this system, O_3 inhibition of the electron transport of photosystem I and photosystem II without uncoupling photophosphorylation can

Table 22.   Ozone effects on apparent photosynthesis.

	Age days	O_3 conc pphm	Duration min	Apparent photosynthesis % of control	Leaf injury %
Bean	51	50	80	65	3
Chard	78	60	70	55	3
Corn	49	50	90	68	3
Cauliflower	77	75	50	38	0
Potato	39	60	60	50	3
Tomato	72	60	60	57	1

Data from Hill and Littlefield (1969). Reprinted with permission from Environ. Sci. Technol., 3, 52-6. Copyright by the American Chemical Society.

be demonstrated. Photosystem I is somewhat more sensitive to O_3 than is photosystem II. Chloroplasts prepared from O_3-treated spinach have decreased photosystem II activity and also decreased B-carotene content (Chang and Heggestad 1974).

The effects of PAN on plant metabolism have been extensively studied and reviewed (Verkroost 1974; Ziegler 1973b). PAN exposure of spinach chloroplasts can inhibit photosystem I and photosystem II without uncoupling photophosphorylation (Coulson and Heath 1975). The inhibition is irreversible and is stimulated by light in aged chloroplasts.

Nitrogen dioxide inhibits apparent photosynthesis, dark respiration and evolution of CO_2 into CO_2-free air by bean leaves over a wide range of environmental conditions (Srivastava et al. 1975b). These effects are due to inhibiting factors within the leaves and not to interference with CO_2 diffusion into the leaf. The uptake of NO_2 is enhanced by high temperature, low CO_2 concentration and high humidity. Inhibition of apparent photosynthesis and dark respiration is increased by increasing NO_2 concentration and exposure time (Srivastava et al. 1975c). Leaf sensitivity is greatest at the ages when maximum rates of apparent photosynthesis are noted in NO_2-free controls. Transpiration rates are less affected, suggesting that the principal effects of NO_2 on gas exchange are exerted in the leaf mesophyll rather than the stomata.

Nitrogen oxide and NO_2 also affect photosynthesis rates of tomato (Capron and Mansfield 1976). Both gases reduced photosynthesis with reduction from 13.03 mg CO_2 $dm^{-2}h^{-1}$ in controls to 9.40 in 50 pphm NO and 8.84 in 50 pphm NO_2. The pollutants do not interact when mixed.

Gaseous exchange rates respond to the interaction of nitrate nutrition and NO_2 exposure in beans (Srivastava et al. 1975a). Exposure to 30 pphm NO_2

depresses photosynthesis and respiration in leaves of plants at all nitrate levels but inhibition is greatest at zero nitrate. The inhibitory effects of NO_2 on photosynthesis and respiration are correlated with a leaf protein basis of activity.

Sulphur dioxide. There have been many reports of SO_2 or sulphite compound effects on photosynthesis. Destruction of chlorophyll by an irreversible oxidation process is a cause of SO_2 injury (Showman 1972). Photosynthetic oxygen evolution is inhibited much more than apparent photosynthesis by SO_2 or bisulfite ions added to isolated spinach chloroplasts (Silvius et al. 1975). Photosynthesis is more effectively inhibited by SO_2 in bean, and SO_2-sensitive species, than in maize, a relatively SO_2-insensitive species (Sij and Swanson 1974). The initial inhibition of photosynthesis is not due to stomatal closure. There is also an interaction between SO_2 and NO_2 in inhibiting photosynthesis of peas (Bull and Mansfield 1974).

Photosynthetic effects of SO_2 have also been studied extensively at the enzyme and organelle level. There is a decrease in chlorophyll fluorescence induced by sulphite (Arndt 1974). Low concentrations of sulphite stimulate the light reactions of photosynthesis and CO_2-fixation of isolated spinach chloroplasts (Libera et al. 1973), but higher concentrations of sulphite can inhibit both these activities and photophosphorylation. Sulphite is effective in inhibiting the CO_2 fixation of corn leaf tissue (Libera et al. 1975). This inhibition is accompanied by a decrease in malate and an increase in aspartate.

Labeled SO_2 is absorbed by spinach leaves and photoreduced, as indicated by the appearance of labeled sulfides (Silvius et al. 1976). This photoreductive activity is considered to be related to the phytotoxic effects of SO_2 as well as to the production and emission of H_2S from illuminated plants exposed to SO_2.

Fluoride. Fluoride is most effective of six air pollutants (HF, O_3, NO_2, NO, SO_2 and Cl_2) in inhibiting photosynthesis of barley and alfalfa (Bennett and Hill 1973).

Hydrogen sulfide. Concentrations of H_2S resulting in decreased photosynthesis and respiration and increased water loss have been established for spinach leaves (Oliva and Steubing 1976).

Trace elements. Net photosynthesis and growth of corn and sunflower plants is decreased most by Ti, followed by Cd, Ni and Pb in hydroponic culture experiments (Carlson et al. 1975). Stomatal opening and transpiration measurements indicate that stomatal movement may be an important factor in determining this effect on net photosynthesis. In addition, Cd has an inhibitory effect on

water uptake in corn, and Ti reduces photosynthesis more than can be attributed
to stomatal closure alone. Inhibitions of photosynthesis rates and stomatal
function by heavy metal trace elements have been reported for soybean, corn and
sunflower (Bazzaz et al. 1974a,b,c). Lead also affects photosynthesis of woody
plants (Rolfe and Bazzaz 1975). The trace elements vary in effectiveness in
inhibiting chlorophyll fluorescence (Arndt 1974). The inhibiting effect of Cd
on chloroplast activity has been described (Bazzaz and Govindjee 1974). Cadmium
strongly inhibits photosystem II but not photosystem I activity, probably at the
O_2 evolving site, and thus inhibition is reversible. Cadmium causes a decrease
in total chlorophyll and in chlorophyll a/chlorophyll b ratio.

6.4 Non-photosynthetic Metabolism and Chemical Constituents

Oxidants. Non-photosynthetic metabolism and chemical composition are also
considerably influenced by O_3. Exposure of bean plants to O_3 induces decreases
in ribonucleic acid and protein levels in primary leaves as well as corresponding increases in the level of ribonuclease (Craker and Starbuck 1972). Leaves
visibly injured by O_3 accumulate free amino acids (Tomlinson and Rich 1967).
Sub-threshold treatments result in increased alanine and γ-amino-n-butyric acid,
and decreased glutamic acid. These effects are considered to be due to sulfhydryl oxidation and increased membrane permeability. Corn pollen exposure to
O_3 provides evidence for the autolysis of structural glyco-proteins and stimulation of amino acid synthesis (Mumford et al. 1972). As little as three pphm
O_3 results in a 50% increase in free amino acids. Higher O_3 concentrations
further increase amino acid and peptide levels and inhibit pollen germination.
Lipid peroxidation does not occur until O_3-induced injury symptoms develop
(Tomlinson and Rich 1970a); nor are unsaturated fatty acids selectively
destroyed in bean leaves exposed to O_3.

Several species exposed to NO_2 form nitrate and nitrite in the leaves,
followed by their reduction to ammonia and amino acids, with ultimately higher
protein in the leaf (Zeevaart 1976). Leaf necrosis induced by NO_2 can be
observed and is attributed, in most species, to acidification as a result of
the reaction of NO_2 with water.

Studies of the effect of nitrate nutrition on NO_2 response in beans show
that the greatest NO_2 uptake is at low nitrate levels (Srivastava et al. 1975a).
Uptake of NO_2 is higher when urea is used instead of nitrate. Uptake of NO_2 is
consistent with an assimilation pathway via nitrate and nitrite reduction.

A number of studies have compared the composition of oxidant exposed and
non-exposed plants at various times after exposure. A thorough analytical
study shows that O_3 exposure results in decreased, then increased soluble sugar

levels, decreased glycolysis, amino acid and protein accumulation, and increased levels of phenolic metabolism enzymes (Tingey 1974). Root growth is reduced more than top growth.

Caffeic acid concentration increases in bean leaves exposed to polluted ambient air compared to filtered air (Howell 1970). Injured leaves are reddish-brown but no anthocyanins can be detected. The chemical nature of the pigments formed in bean leaves as a result of O_3 exposure has been studied (Howell and Kremer 1973). Dark-coloured polymers are formed apparently as a result of impairment of membrane integrity which allows phenolase activity on phenolic compounds which are oxidized to quinones. The quinones produced polymerize with amino acids and proteins, possibly reducing nutritive value of leaves. There are many effects of pollutants on phenolic compounds (Howell 1974). The red colouration appearing on the leaves of dock plants after O_3 exposure is due to anthocyanin formation (Koukol and Dugger 1967). The anthocyanin is a cyanidin glycoside.

'Pinto' bean plants exposed to O_3 contain disulfides while controls contain none (Tomlinson and Rich 1970b). There is no corresponding decrease in sulf-hydryl content. Tomato and bean plants exposed to O_3 have increased rates of ethylene production, reaching physiologically active concentrations in two to four hr in a closed system (Craker 1971). Some responses of plant tissue to O_3 could thus be due to ethylene production by injured tissue.

Ascorbic acid analysis of petunia leaves reveals a parallel of ascorbic acid content and O_3 sensitivity, that is, younger leaves had both highest ascorbic acid content and least O_3 sensitivity (Hanson et al. 1971). Cultivar differences in O_3 sensitivity are also reflected in different ascorbic acid levels. Mature leaves also increase in both ascorbic acid and O_3 insensitivity as the flowering stage is approached.

Sterols and sterol derivatives, components of membranes, are changed by O_3 exposure (Tomlinson and Rich 1971). Acylated sterol glycosides and sterol glycosides increase consistently as the free steroids decrease in 'Pinto' bean leaves. Many fatty acid components increase with major increases in linolenic acid.

Sulphur dioxide. Sulphite and SO_2 have considerable effects on plant res-piration (Verkroost 1974; Ziegler 1973b). Sulphite is an effective inhibitor of ATP formation by plant mitochondria prepared from etiolated tissues of bean and corn (Ballantyne 1973). Sulphite oxidation by mitochondrial preparations from green or etiolated tissues has been compared (Ballantyne 1977). Mitochondria from both tissues effectively oxidize sulphite with the etiolated tissue more sensitive to sulphite. If SO_2 becomes sulphate in the tissue, the sulphite toxicity may be alleviated by this oxidation.

Exposure to SO_2 might also involve amino acid metabolism. Photosynthesis is reversibly inhibited by exposing plants to SO_2 because a-hydroxysulphonate is formed in the leaves of SO_2-exposed wheat plants. It is suggested that the glycolytic acid pathway may be suppressed, causing a decrease in the photosynthetic biosynthesis of serine (Tanaka et al. 1972a,b). Glyoxal-bisulphite is formed in SO_2-treated rice plants (Tanaka et al. 1972a). Accumulation of sulphite at the reaction site could considerably reduce CO_2 assimilation (Tanaka et al. 1974). Accumulation of labeled glycolic acid (from photosynthetically fixed CO_2) occurs in the leaves of barley plants exposed to SO_2 (Spedding and Thomas 1973). The relative importance of inorganic and organic SO_2 derivatives in determining toxicity of SO_2 has been discussed (Pahlich 1975).

Nitrogen fixation is more susceptible to sulfite injury than is photosynthesis in a lichen and a blue-green alga (Hallgren and Huss 1975). Treatments at pH 5.8 are more inhibitory than at higher pH values and the inhibition of nitrogenase activity and CO_2-fixation is less in darkness and increases with increasing light intensity. Both processes recover after removal of the sulfite solution.

Exposure of pea plants to SO_2 increases inorganic S, reduces buffer capacity, and stimulates glutamic dehydrogenase (Jager and Klein 1977) while K, Ca, P and N fractions are not affected. Labeled S from SO_2 is incorporated into thioglycosides (Spaleny et al. 1965), breakdown products of which have bactericidal, fungicidal or herbicidal effects. The mechanism of SO_2 toxicity may relate to this effect.

Fluoride. The effects of F on plant metabolism have been extensively described (McCune and Weinstein 1971; Verkroost 1974; Ziegler 1973b; and others). Respiration is stimulated in bean and gladiolus by prolonged F exposure (McNulty and Newman 1957). Oxygen uptake by bean plants is accelerated by F concentrations below those inducing foliar symptoms (Applegate and Adams 1960b). This respiration effect is considered an example of "invisible injury" (Applegate and Adams 1960c).

The respiratory pathway is related to gladiolus HF sensitivity (Ross et al. 1968). The primary F injury is considered to involve the inhibition of enolase, with insensitive cultivars more dependent than sensitive cultivars on the pentose phosphate respiratory pathway and less on the glycolytic pathway. Mitochondria prepared from HF-treated soybean plants have higher respiration rates and ATP-ase levels but lower ADP/O ratios (Miller and Miller 1974). Fluoride also increases osmotically-induced swelling and protein leakage from mitochondria prepared from corn. Fluoride-induced potato phosphoglucomutase inhibition depends on the amount of Mg^{++} present (De Maura et al. 1973). There are inhibiting effects of F on protein synthesis as well as the stimulating effect of F on RNA-ase activity in Lens roots (Pilet 1969, 1970) and in corn roots (Chang 1970, 1973).

Corn seedlings can be used to demonstrate that F-induced growth retardation is associated with changes in protein formation and that F-induced aging is associated with changes at the site of protein synthesis (Chang 1973). With F treatment, total and ribosomal RNA decreases, ribosomal components are altered, and ribosomal distribution shifts from polysomes to smaller particles. ATP accumulates, ribonuclease activity in roots is activated, and phytase activity in seeds is inhibited.

Fluoride affects the chemical composition of bean and tomato plants (Weinstein 1961). Free sugar content of leaves and chlorophyll formation are reduced in both species by F exposure while RNA-phosphorus is reduced in tomato but not in bean. The length of exposure is an important determinant of metabolite concentration.

The distribution of P compounds is not affected by HF exposure of bean seedlings even though considerable F is accumulated (Pack and Wilson 1967). Apparently these plants are able to tolerate F accumulation without inhibition of P metabolism enzymes. Studies of the nature of F-induced chlorosis in bean leaves indicate that NaF prevents chlorophyll a and b and protochlorophyll accumulation in previously etiolated tissue, but transformations of these pigments are not affected (McNulty and Newman 1961). Only early stages of pigment synthesis appear to be affected along with degradation of the chloroplast structure. As for O_3, F exposure can induce an increase in phenolic contents of tissue (Yee-Meiler 1974).

Trace elements. Cadmium and Pb treatments inhibit soybean growth and N-fixing activity in nodules containing Rhizobium japonicum (Huang et al. 1974). Lead may inhibit or stimulate maize mitochondrial respiration depending on the concentration and anions present (Koeppe and Miller 1970). The inhibition appears to be reversible.

Pesticides. Some pesticides may affect photosynthesis and respiration and may frequently hinder eventual fruit production and quality (Ayers and Barden 1975). Interactions of mixtures, as well as multiple and concentrate application methods, need evaluation.

Salinity. Nitrogen fixation by pea root nodules decreases as nutrient solution salt concentration increases (Minchin and Pate 1975). Accumulation of salts on nodule surfaces is associated with marked inhibition of fixation.

6.5 Anatomical and Cytological Effects

Early studies of anatomical effects of pollutants were conducted with the
light microscope. For example, information is available on the histological
responses of leaves to SO_2 and HF, including photographs of effects on apricot,
apple and tomato leaves (Solberg and Adams 1956). Anatomical effects of both
pollutants are similar and consist of collapse of spongy mesophyll and lower
epidermis followed by distortion and chloroplast disruption in the palisade
cells. Histological responses to O_3 have also been described (Ledbetter et al.
1959). More recently, the electron microscope has been utilized for ultrastructure studies of injured cells. Electron micrographs are available showing
O_3, PAN, NO_2, SO_2, HF and ethylene effects (Thomson 1975). Characteristic
alterations in cell ultrastructure occur with pollutant exposure, in some cases
prior to the appearance of visible symptoms. There is usually a sequential
progression of changes which includes membrane alterations, loss of organiza-
tion and compartmentation, and frequently the appearance of crystalloids.
There may also be injury to the genetic components of the cells resulting in
impaired reproduction.

Oxidants. Ozone effects on 'Pinto' bean leaf ultrastructure occur in two
phases, the first a granulation and increase in electron density, and the second
a disruption of membranes and organelles, sometimes including the grana within
the chloroplasts (Thomson et al. 1966). The effects are similar to those induced
by PAN and both are considered to result from the oxidizing properties of the
molecules. Ozone-induced changes in the ultrastructure of bean leaf mesophyll
cells include early changes in the cell membranes along with the occurrence of
crystalline bodies in the chloroplasts (Thomson et al. 1974). Progressive O_3
injury from cell to cell can be noted in electron micrographs of pea leaf mesophyll cells with wide differences in injury in adjacent cells (Fig. 41).
Studies of the interaction of O_3 and benomyl, an antiozonant, at the ultrastructural level indicate that the chloroplasts of bean leaves treated with
benomyl and exposed to O_3 remain intact and are similar to control leaf chloroplasts (Rufner et al. 1975). Benomyl plus O_3-treated leaves have a higher
chlorophyll content than O_3-treated leaves.

Studies of O_3-induced chromosomal injury in broad bean indicate that O_3
causes chromosomal injury in root tips (Fetner 1958). Exposure of plants to
relatively high concentrations of O_3 followed by examination of chromosomes
in bud tissue demonstrates that chromosomal injury consists of stickiness,
bridges, fragments and micronuclei (Janakiraman and Harney 1976). Chromosomes
are more susceptible to O_3 during early stages of meiosis than during later
stages. However, reciprocal cross pollinations between O_3-exposed and unexposed

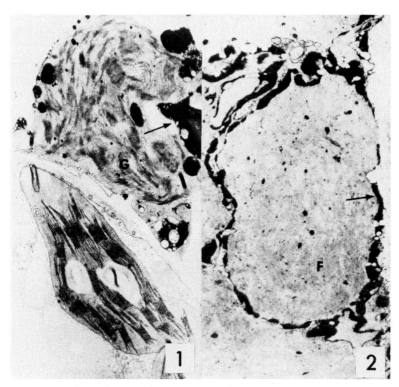

Fig. 41. Ozone effects on pea leaf ultrastructure. (1) Mesophyll tissue of a
young leaf showing two adjacent cells. One cell is uninjured and
the chloroplast shows well defined inner membrane systems. The
injured cell is characterized by electron dense precipitates at the
periphery (→), and loss of integrity of inner membrane systems.
G = grana. (2) Mesophyll tissue of an old leaf showing the chloroplasts of a severely injured cell. The stroma is fibrillar and
there is an accumulation of electron dense precipitate at the
envelope (→). F = fibrillar stroma (From Mitchell, Ormrod and
Dietrich, unpublished).

tomato plants result in normal fruit development with viable seeds (Gentile
et al. 1971).

Nitrogen dioxide, sulphur dioxide. Exposure of beans to either NO_2 or SO_2
results in reversible swelling of the thylakoids within the chloroplasts without

extra-chloroplastic injury (Wellburn et al. 1972). The swelling may be
associated with reduction in photosynthesis rates caused by these pollutants.
Effects of SO_2 on bean chloroplasts have also been examined by Godzik and
Sassen (1974).

Fluoride. The only histological change observed away from the necrotic
areas of geranium leaves exposed to HF is the formation of tyloses in xylem of
veins and petioles, indicating the F injury may influence plant function some
distance from the site of visible injury (Poovaiah and Wiebe 1969). Changes in
the ultrastructure of soybean leaves following HF exposure have also been
described (Wei and Miller 1972).

The site of F accumulation in orange leaves can be studied using a frac-
tionation procedure involving organic solvents (Chang and Thompson 1966). The
chloroplast fraction has the highest F accumulation.

Fluoride exposure of corn plants at about three mg/m^3 followed by cytological examination results in evidence that HF is a mutagenic agent (Mohamed 1970).
Chromosome aberrations include asynaptic regions, translocations, inversions,
bridges and fragmentation. Similarly, exposure of tomato seeds to HF gas
followed by seeding results in some plants with abnormal numbers of cotyledons,
deformed cotyledons, fasciated petioles, absence of plumules, dwarfism, and
other deformities (Mohamed 1968). Chromosomal aberrations are noted in the
offspring of treated plants with generally more aberrations with an increase
in HF exposure duration. The role of HF as a mutagenic agent is attributed to
blocking the replication of DNA.

Other air pollutants. Leaf mesophyll cells of corn exposed to pollutants,
including SO_2 and CO as well as cement dust from a cement factory, have breaks
in cell membranes along with dense accumulations and vacuole-like structures
in the cytoplasm (Cireli 1976).

Pesticides. Phenolic compounds (either pesticides or degradation
products of pesticides) applied to broad bean to observe effects on meiosis
induce abnormal pollen mother cells but do not affect pollen viability (Amer
and Ali 1968). The meiotic irregularities are stickiness, lagging chromosomes,
anaphase bridges, and fragmentation. P-nitrophenol is the most toxic compound
of those tested. Studies of the effects of sevin on meiosis in broad bean
indicated that increasing the number of sevin sprays increases the number
of abnormal pollen mother cells but does not affect pollen fertility (Amer
and Farah 1968). Stickiness, lagging chromosomes, and other irregularities
occur in meiosis. Polyploidy can be induced by sevin treatments.

CHAPTER 7

GENETICS AND PLANT BREEDING

Knowledge of relative insensitivity of horticultural plant species and cultivars to pollutants is necessary for the recommendation of specific plant types for areas with high frequency of plant-injuring concentrations of pollutants. Such information is also useful for the selection of genetic lines or accessions for breeding programmes (Gabelman 1970). Research on the selection and breeding of plants that are genetically less sensitive to pollutants has been reviewed (Ryder 1973). Considering the potential importance of the plant breeding approach to protection of horticultural plants against pollutants, there have been surprisingly few plant breeding programmes undertaken.

7.1 Cultivar Testing

Evaluation of cultivar reactions to pollutants is a widely used method in the search for genetically resistant plant material. Cultivars may be exposed to pollutants in controlled environment chambers to evaluate differential responses. Alternatively, responses to pollutants in the field or greenhouse may be evaluated. Sites at which pollutants are known to occur can be used as test locations. Plants shown to be insensitive to the range of toxicants occurring in many urban and industrial locations may have more likelihood of success in comparable locations elsewhere.

Cultivar screening for visible pollutant injury to foliage is appropriate for crops which have foliar appearance requirements for marketing or acceptability. Species that produce fruit or roots for marketing should have the pollutant effect on that component of growth measured, rather than using foliar injury criteria as the sole basis for cultivar selection. In fact, foliar injury and yield reductions due to O_3 may not be well correlated in tomato cultivars (Oshima et al. 1977a).

Some innovations in cultivar testing have been proposed. For example, the high humidity in greenhouses cooled with fan and pad systems would make such facilities useful for pollution sensitivity studies (Klingaman and Link 1975). The higher humidity predisposes plants to pollutant sensitivity. Greenhouses in areas of high ambient pollution could be used to determine a practical level of pollution tolerance in selection programmes. Another innovation would be the use of chemical protectants to determine cultivar differences in pollutant sensitivity. Yields of plants treated with appropriate protectants could be compared with untreated ones in areas with ambient pollution.

7.2 Genetic Studies

Oxidants. In onion, the mechanism of insensitivity to O_3 injury is that of genetically-controlled stomatal closure (Engle and Gabelman 1966). The guard cells of insensitive cultivars are injured, collapsing and causing the stomates to close, thus lending protection to the plants. The guard cells remain turgid and stomates open in susceptible cultivars. None of a large number of bean lines has complete immunity to O_3 toxicity, suggesting a quantitative basis for insensitivity (Prasad et al. 1970).

The degree of inheritance of sweet corn insensitivity to O_3-induced leaf injury under field conditions in California, U.S.A., has been determined (Cameron 1975). First generation progeny of sensitive X insensitive inbreds are susceptible to injury. Second generation progeny segregate quantitatively with the distribution toward low injury. Insensitive sublines are stable after three generations of selection and inbreeding while segregation continues in susceptible selections.

The inheritance of photochemical pollutant insensitivity in petunia has been studied using all possible combinations of crosses among seven inbred lines of a pink-flowered multiflora petunia differing widely in sensitivity (Hanson et al. 1976). The genes for insensitivity act primarily in an additive manner, with indications of partial dominance. Hybrids sensitive to one oxidant are not necessarily sensitive to others. Two general mechanisms have been proposed to explain the genetic basis for O_3 insensitivity in petunia (Thorne and Hanson 1976). The gas exchange potential of the insensitive cultivars could be limited, thereby reducing O_3 diffusion into the leaf, and/or some biochemical reaction could neutralize the O_3 after it diffuses into the leaf. The O_3 responses of a seven-parent diallel set of petunia hybrids were studied by determining gaseous exchange rates and other parameters. The research indicated that sensitivity is highly correlated with O_3 and water vapour diffusion resistance, but not with ascorbic acid concentration or shoot-root ratio.

Sulphur dioxide. The weedy winter annual, Geranium carolinianum L., provides evidence for infra-specific differences in SO_2 sensitivity depending on the location of sampling (Taylor and Murdy 1975). Insensitive responses are obtained only in plants collected from an area exposed over a 31-year period to SO_2 pollution, suggesting the evolution of genetic insensitivity to SO_2 in this species.

PART II

SPECIFIC RESPONSES TO POLLUTANTS

CHAPTER 8

VEGETABLE CROPS

8.1 Bean

The species Phaseolus vulgaris includes horticultural cultivars used for canning, freezing and fresh marketing, as well as agronomic cultivars used for production of dry white or Navy beans. Some cultivars within both groups have substantial sensitivity to pollutants. In particular, the cultivar 'Pinto' is widely studied because it is very sensitive to O_3, SO_2, NO_2 and PAN. It is used as an indicator plant particularly for its O_3 and PAN sensitivity. There have been many research projects on bean responses to pollutants. These will be reviewed, whether involving horticultural or agronomic cultivars, because of the many similarities in the responses of all bean cultivars.

Oxidants. The disorder of bean called bronzing has been noted for many years in oxidant-prone production areas. The upper leaf surface becomes bleached or develops brown spots on plants exposed to controlled O_3 treatments (Middleton 1956). The environment under which the plants are grown determines whether the leaves bleach or develop reddish brown to dark brown punctate spots, all symptoms being limited to the upper leaf surface. Growth reductions, without visible injury symptoms, are also found when beans are exposed to ozonated hexene (Todd and Garber 1958). Leaves are smaller and senesce sooner. Ozone or ozonated hexene causes a four-fold increase in respiration rate of 'Pinto' bean leaves while photosynthesis rate is reduced, with the changes proportional to development of visible injury symptoms (Todd 1958). Exposure of young 'Pinto' bean plants to five pphm O_3 results in premature senescence of unifoliate leaves and decreases in leaf size and weight (Engle and Gabelman 1966).

The onset of oxidant-incited bronzing on beans in the field has been described (Haas 1970). Major variations in bronzing incidence are found in apparently uniform field locations. Most bronzing occurs about 10 days after the full bloom stage and about 10 to 14 days before normal rapid defoliation is expected. Carbohydrate depletion in the leaves is suggested as a factor predisposing them to oxidant-incited chlorosis, necrosis and defoliation. Metabolic changes associated with injury development have also been described

(Pell and Brennan 1973). The chemical nature of the dark-coloured pigments formed in bean leaves as a result of O_3 exposure has been characterized (Howell and Kremer 1973). Polymers are apparently formed as a result of phenolase activity on phenolic compounds. The quinones produced polymerize with amino acids and proteins, possibly reducing the nutritive value of leaves.

There have been several studies of cultivar response to O_3, and there is a wide range of sensitivity as in many other species. Several bean strains from a world collection, as well as North American cultivars, have a range of symptom expression and injury severity and there is a relationship between O_3 insensitivity and <u>Rhizoctonia</u> damping-off disease resistance (Prasad et al. 1970).

Snap bean cultivars as a group are much less sensitive to O_3 than white or Navy bean cultivars (Davis and Kress 1974). The green snap bean cultivars 'Astro', 'Harvester' and 'Bush Blue Lake 274' are moderately sensitive, while 'Provider', 'Stringless Black Valentine', 'Tendercrop' and 'Eagle' are moderately insensitive to controlled O_3 exposures. Also, white flower colours tend to be associated with sensitivity and lavender with insensitivity (Table 23).

Table 23. Ozone sensitivity of unifoliate leaves of bean cultivars.[z]

Cultivar	Flower colour	% Tissue injured	Sensitivity rating
Sanilac	White	33	Very sensitive
Pinto III	White	22	Sensitive
Tempo	White (yellow hue)	21	Sensitive
Astro	White	11	Intermediate
Bush Blue Lake 274	White	9	Intermediate
Harvester	White (yellow hue)	6	Intermediate
Eagle	White	4	Moderately insensitive
Stringless Black Valentine	Lavender	4	Moderately insensitive
Provider	Lavender	2	Moderately insensitive
Tendercrop	Light Lavender	1	Moderately insensitive

[z] Exposed to 25 pphm O_3 for 4 hr at $21^\circ C$, 75% RH and 25 klux, one month from seeding.

Data from Davis and Kress (1974)

Oxidant responses of bean cultivars observed in the field showed that 'Tempo', 'White Half Runner', and 'Bush Blue Lake' are most sensitive and 'Long Tendergreen' and 'Gold Crop' least sensitive of 15 cultivars evaluated (Brennan and Rhoads 1976). Development of leaf chlorotic stipple and bronzing is correlated with oxidant concentrations greater than 3.8 pphm for six or seven hr, the day prior to injury severity observation.

A yield experiment with 'Tendergreen' snap beans grown in open-top chambers with or without charcoal filtration of incoming air at Yonkers, New York, U.S.A. indicated that the weight and number of marketable pods is reduced by 26 and 24% respectively, by unfiltered air compared with ambient air from which 60 to 70% of the ambient photochemical oxidants have been removed (Table 24).

Table 24. Ambient oxidant effect on the yield and dry matter of 'Tendergreen' bean plants.

	C-filtered air	Unfiltered air
Pods harvested (no.)	88.4[**]	70.0
Marketable pods (no.)	70.4[**]	53.5
Unmarketable pods (no.)	18.0	16.5
Harvested pods fresh wt (g)	424.9[**]	320.9
Marketable pods fresh wt (g)	390.6[**]	288.6
Unmarketable pods fresh wt (g)	34.3	32.3
Tops dry wt (g)	80.9[**]	65.6
Leaves dry wt (g)	41.5[*]	35.2
Stems dry wt (g)	39.4[*]	30.4

Differences between values from C-filtered and unfiltered air significant at 5% level (*) and 1% level (**).

Data from MacLean and Schneider (1976). Reproduced from JOURNAL OF ENVIRON-MENTAL QUALITY 5, 75-8 by permission of the American Society of Agronomy, Crop Science Society of America and Soil Science Society of America.

Additional facets of bean responses to oxidant have also been researched, for example, studies of the leaf age effect on oxidant-induced injury (Middleton et al. 1950). After three hr controlled exposure to oxidant, the leaf injury indices for successive leaves from the base to the apex of bean plants are 7.8, 6.7, 6.1, 4.3, 2.5 and 0, indicating a much reduced sensitivity in young, compared to old leaves.

Snap beans can be used to demonstrate that plants exposed to low O_3 have reduced injury when later exposed to a higher O_3 level (Runeckles and Rosen 1974).

Mixtures of O_3 and SO_2 have different effects than the separate gases in causing injury symptoms on beans (Hofstra and Ormrod 1977). Visible injury symptoms are different for the combined and separate gases (Fig. 42). The response to the mixture is similar for a range of seven to 75 pphm SO_2, combined with 15 pphm O_3. Also, injury development is retarded in the mixture compared to O_3 alone. This delayed injury development in the mixture

Fig. 42. O_3 injury symptoms on bean unifoliate leaf (left) compared with
O_3 + SO_2 injury. The plants were exposed to 15 pphm O_3 for 10
days or 15 pphm O_3 + 15 pphm SO_2 for 10 days (Author's photo).

correlates well with the results of a sulphate nutrition study in which increased
S nutrition protected bean plants from O_3 injury, especially at lower temperatures (Fig. 43). Combined and separate O_3 and SO_2 exposures of bean have also
been carried out by Jacobson and Colavito (1976) who demonstrated that injury
is greater from O_3 than from SO_2 when supplied separately. The two gases are
either antagonistic or behave independently when combined, depending on their
concentrations.

Environment effects. Beans are widely used to show that water stress is
an important determinant of oxidant injury. 'Pinto' bean plants are protected
against O_3 or irradiated auto exhaust injury by withholding water prior to
exposure (Seidman et al. 1965). Factors controlling water vapour loss from
bean leaves also regulate O_3 removal from the air (Rich et al. 1970). Stomatal
opening is thus considered to be an important factor controlling O_3 uptake.
Stomatal resistance responses of 'Pinto' bean plants exposed to O_3 at different
soil, water and atmospheric humidity regimes have been measured (Rich and Turner
1972; Markowski and Grzesiak 1974). Stomata of soil water stressed, but not
wilted, leaves close quickly on exposure to O_3, while those of unstressed plants
close more slowly. Stomata on plants in a dry atmosphere close in the presence
of O_3 but open again after O_3 treatment is terminated. Stomata are unaffected
by O_3 in a moist atmosphere.

118

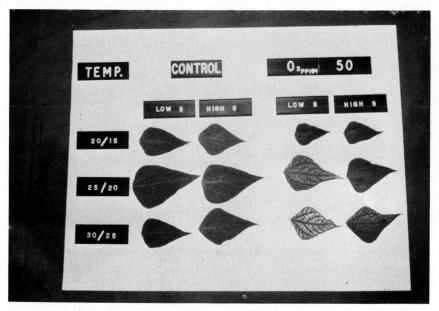

Fig. 43. Effects of growth temperature and sulphur nutrition on the response of
the terminal leaflet of the first trifoliate of bean to O_3 fumigation
at 50 pphm for 2 hr (Adedipe et al. 1972b. Reproduced by permission
of the National Research Council of Canada from the Canadian Journal
of Botany, 50, pp. 1789-92).

Some other 'Pinto' bean responses to O_3 have been characterized (Dunning
et al. 1974). Plants are more sensitive when grown and exposed at 28 compared
to 20°C. Exposure duration interacts with temperature response in affecting
sensitivity. There is more foliar injury at low K nutrition but no relationship
to reducing sugars or sucrose content of the leaves. Other factors affecting
'Pinto' bean responses to O_3 are the light intensity, temperature and humidity
regimes (Dunning and Heck 1973, 1977).

Protection. Studies of protection of beans indicate that oxidant injury
to plants sprayed with 0.01 M K ascorbate is about 40% of that for control
plants (Freebairn and Taylor 1960). Feeding of K or Ca ascorbate through the
roots also protects bean plants against oxidant injury (Freebairn 1963). Other
approaches to chemical protection against O_3 injury include antioxidants, thiols
and sulfhydryl reagents (Dass and Weaver 1963). Ascorbic acid is less effective
than Ni-n-dibutyl dithiocarbamate. Abscisic acid treatment of primary leaves

of bean plants reduces O_3 injury (Fletcher et al. 1972). Bean plants are protected against ozonated gasoline or hexene-1 by sprays or dusts of zineb, maneb,
thiram or ferbam, but not Bordeaux mixture, dichlone or chloranil (Kendrick
et al. 1954). Protection of bean plants in both field and laboratory studies
is directly proportional to the concentration of effective chemicals used
(zineb, maneb, thiram and ferbam) and the action is localized, not systemic
(Kendrick et al. 1962). Young 'Pinto III' bean plants exposed to O_3 have less
injury if treated with the systemic fungicide triarimol as either foliar sprays
or soil applications (Seem et al. 1972). 'Pinto III' beans are also protected
against O_3 injury by foliar or soil applications of thiophanate ethyl (Seem et
al. 1973). Nearly complete protection results from a soil application rate of
200 mg/g soil (Table 25).

Table 25. Ozone protection by application of thiophanate ethyl to 'Pinto III'
bean plants.[y]

	Active ingredient conc μg/g or ml	Primary leaf area injured %
Foliar spray	0	24.4a[z]
	100	20.6a
	500	7.8b
	1000	1.6b
Soil amendment	0	24.7a
	100	6.3b
	200	0.6b
	500	0.6b

[y] Exposed to 25-35 pphm O_3 for 4 hr.
[z] Mean separation by Duncan's modified least significant difference test,
1% level.
Data from Seem et al. (1973). Reproduced from JOURNAL OF ENVIRONMENTAL QUALITY,
2, 266-8, by permission of the American Society of Agronomy, Crop Science
Society of America and Soil Science Society of America.

Beans were used in much of the research on the use of the systemic fungicide benomyl as an O_3 protectant. Soil application of benzimidazole or benomyl
protects 'Pinto' beans against O_3 (Pellissier et al. 1972a, b). Benomyl sprays
each week on 'Tempo' and 'Pinto III' beans reduce oxidant-induced leaf injury
by 75 - 80% but only in 'Tempo' is there increased growth and yield (Manning et
al. 1974). Unsprayed plants have extensive oxidant injury, mature faster, and
set fruit earlier than sprayed plants, which have only slight visible injury.
Benomyl soil amendments provide only short-term O_3 protection (Manning et al.

1973c). The technique of using benomyl foliar sprays to determine ambient O_3 effects on yield under field conditions has been used for yield studies of 'Tempo' bean, an O_3 sensitive cultivar (Manning and Feder 1976). Benomyl-sprayed plants have about 30% higher yield than unsprayed.

Beans growing in soil containing 1.2 to 9.5 ppm of carbathiin (carboxin) or other 1,4-oxathiin derivatives are protected against O_3 injury (Rich et al. 1974; Manning and Vardaro 1973). Carbathiin sprays vary in effectiveness Curtis et al. 1975; Manning and Vardaro 1973). Benomyl and carboxin soil drenches combined with contact nematicides result in 'Pinto' beans resistant to O_3 injury (Miller et al. 1976) with a combination of benomyl and fensulfothion inducing O_3 insensitivity more rapidly than benomyl alone. Beans are protected against O_3 injury by foliar diphenylamine dusts (Gilbert et al. 1975). Application of N-6-benzyladenine or kinetin to beans modifies the O_3 response but gives little protection against O_3 injury (Runeckles and Resh 1975b). Chlorophyll loss is reduced but weight loss induced by 10 pphm O_3 is not prevented.

Protection of beans against O_3 injury can also be obtained by spraying on water solutions of a paraffinic wax emulsion (Pellissier et al. 1972c). Protection occurs only on sprayed leaves or leaf parts. Complete protection requires the spraying of both leaf surfaces.

Peroxyacetyl nitrate. Injury symptoms resulting from PAN exposure of beans include abaxial leaf surface bronzing and glazing, adaxial leaf surface stipple or fleck, and bifacial necrosis (Starkey et al. 1976). 'Provider' is the most sensitive cultivar of 10 tested and it is suggested for use as a PAN indicator because it is apparently insensitive to O_3 while sensitive to PAN (Table 26).

There is a requirement for a post-exposure light period for PAN injury development (Taylor et al. 1961). A short exposure of 'Pinto' bean plants to PAN requires a two to four hr light period after exposure for injury to occur. Protection against PAN or O_3 + PAN injury is obtained by benomyl soil drenches on 20-day-old 'Pinto' bean plants but is ineffective on plants five days younger (Pell 1976).

Nitrogen Oxides. The supply of NO_3 during growth affects the onset and intensity of NO_2-induced injury to bean leaves (Srivastava et al. 1975a). Yellow and brown patches and marginal necrosis are greatest at low NO_3, and injury is decreased by increasing NO_3 supply during growth.

Sulphur Dioxide. Injury from SO_2 on beans is related to soil water stress (Markowski et al. 1974). There is little or no injury after exposure under drought conditions while exposure under optimum soil water conditions results

Table 26. Ozone and PAN susceptibility of bean cultivars at the secondary leaf
stage.

	O_3	PAN^y
1.^{z}	Sanilac	Provider
2.	Pinto III	Astro
3.	Tempo	Harvester
4.	Astro	Sanilac
5.	Bush Blue Lake 274	Bush Blue Lake 274
6.	Harvester	Tendercrop
7.	Eagle	Pinto III
8.	Stringless Black Valentine	Stringless Black Valentine
9.	Provider	Tempo
10.	Tendercrop	Eagle

^{y} Based on bifacial necrosis.

^{z} Order of decreasing susceptibility (i.e. 1 = most sensitive).

Data from Starkey et al. (1976)

in reduced yield, height and chlorophyll content. Sulphate S analyses can be
used to show that a lower hydration of tissues is associated with lower SO_2
injury. Bean leaves exposed to SO_2 have a wider opening of stomata when plants
are growing under drought conditions but there is no measurable effect on
diffusive resistance, which remains high (Schramel 1975). The protective
effect of drought against SO_2 toxicity is thus related to a high diffusive
resistance of leaves which is not associated with the influence of SO_2 on
stomatal movement.

Fluoride. Growth of 'Pinto' beans is accelerated in the presence of low
levels of F but tissue F contents over 200 ppm are associated with growth retardation, and over 500 ppm with leaf chlorosis (Treshow and Harner 1968). Bean
pod weights are reduced almost 25% by exposure to 0.6 mg F/m^3 during growth
(MacLean et al. 1977).

Nutrient deficiencies affect the F response of bean plants (Applegate and
Adams 1960a). Plants mildly deficient in P, Fe or K take up more F than plants
deficient in N or Ca. Complex interactions of nutrient deficiency and F effects
occur in respiration. Bean seedlings exposed to HF gas are more sensitive if
grown in N-deficient conditions compared to K, Ca, P and Fe-deficient plants,
even though the latter have higher tissue F concentrations (Adams and Sulzbach
1961).

Generation carryover and seed effects of HF exposure have been studied
using snap beans grown with continuous HF exposure (Pack 1971a,b). Subsequent
generations are less vigorous and have more leaf abnormalities than progeny of

control plants (Pack 1971a). Somewhat fewer fruit and fewer seed per fruit are produced on bean plants exposed to HF as low as 2.1 mg F/m^3 from seeding to harvest (Pack 1971b). Affected seed is faded, shriveled, lighter in weight and lower in starch content than control plant seed. These fruiting effects occur without foliage injury.

Bush beans exposed to atmospheric F have higher respiration rates beginning after eight days at 30 ppb F, with older leaves having a greater increase in respiration (McNulty and Newman 1957). Fluoride affects several metabolic constituents of bean (Weinstein 1961).

Acid Rain. Simulated acid rain can have deleterious effects on 'Pinto' bean plants (Ferenbaugh 1976). Sprays of H_2SO_4 solutions of pH less than three reduce height and cause necrotic and wrinkled leaves, excessive adventitious bud development, and premature abscission of leaves.

Trace Elements. Snap beans have been used to test the effects of Cu application for pathogen control (Walsh et al. 1972). Yield reductions occur when Cu concentrations in seedling trifoliate leaves exceed 20 to 30 ppm with severe toxicity at levels in excess of 40 ppm Cu. These tissue levels are associated with application rates of about 130 kg Cu/ha and about 450 kg Cu/ha respectively (Table 27).

Table 27. Germination and yield of snap beans grown on Plainfield loamy sand as affected by rate and carrier of Cu.

Carrier and Cu rate (kg/ha)		Germination (plants/0.3 m row)		Yield (pods)[z]	
		Year 1	Year 2	Year 1	Year 2
				---- kg/ha ----	
Control	0	6.1	8.5	6,160	8,176
$CuSO_4$	18	6.6	8.7	6,563	9,374
	54	6.4	9.1	4,861	8,299
	162	5.7	8.2	3,819	8,042
	486	5.4	7.3	1,008	1,938
Cu (OH)$_2$	15	6.7	8.4	5,846	10,315
	45	6.2	9.0	5,891	7,896
	135	6.1	8.0	5,118	7,011
	405	5.8	8.8	4,659	5,376
Lsd (.10)		0.7	0.6	2,834	1,355

[z] Stand density 5 and 7 plants/0.3 m row in year 1 and 2 respectively. Data from Walsh et al. (1972). Reproduced from JOURNAL OF ENVIRONMENTAL QUALITY, 1, 197-200, by permission of the American Society of Agronomy, Crop Science Society of America and Soil Science Society of America.

Beans have also been used to demonstrate that Cr toxicity in soil occurs only if EDTA is also added (Wallace et al. 1976). In solution EDTA has less effect on Cr toxicity because the Cr is already in solution. Bush bean plants can be used to assess Sn toxicity (Romney et al. 1975). Very little Sn is translocated to shoots and yields are reduced only slightly by 500 ppm Sn in the soil. Addition of $CaCO_3$ to nutrient solutions containing $10^{-3}M$ Sn prevents severe toxicity.

Salt. Beans are relatively sensitive to salt (Hayward and Bernstein 1958). Simulated seashore salt aerosols applied to beans showed that Na and Cl uptake is linear with time and onset of toxicity symptoms corresponds with a leaf tissue level of 2.6% Cl (Williams and Moser 1975). The research was unique in that deposition chambers were used in which aerosols with different sedimentation rates could be obtained. Such aerosols would be found in ambient conditions in seashore areas. Application of simulated saline cooling water mist to beans causes salt toxicity, the degree of which is affected by humidity (McCune et al. 1977). Increase from 50 to 85% relative humidity doubles the injury caused by the mist. Increase in the fraction with particle size greater than 150 μm, increases the toxicity compared to particles between 50 and 150 μm.

8.2 Table Beet

Sulphur dioxide sensitivity is a recognized characteristic of table beets (Jacobson and Hill 1970), but there has been little back-up research. Ozone-induced foliar injury is noticeable at levels as low as 20 pphm for one hr per day (Ogata and Maas 1973). Growth is reduced by longer exposures. Studies of salinity effects on O_3 response of table beets indicate that there is an interaction with higher rooting medium salinity levels resulting in decreased yield but also decreased O_3 injury.

8.3 Broccoli

Broccoli is relatively sensitive to SO_2 (Jacobson and Hill 1970) but this species has had little research study. In an early report, it was noted that O_3 causes conspicuous tumours on broccoli leaves that are especially pronounced on the lower leaf surface (Hill et al. 1961). This species has been included in gas mixtures studies (Tingey et al. 1973d). Plants are quite insensitive to O_3 levels of up to 10 pphm while they are severely injured by 100 pphm SO_2 but not by 50 pphm. Considerable leaf injury occurs after exposure to mixtures of O_3 and SO_2 at concentrations as low as 10 pphm of each gas. Cabbage responds very much like broccoli.

8.4 Carrot

Rated as sensitive to SO_2 (Jacobson and Hill 1970) carrots have only recently had air pollution research attention. Carrot plants exposed intermittently to O_3 develop chlorotic leaves and plants have less root dry matter than controls, but plants are taller and have more leaves in spite of the chlorosis (Table 28). It is estimated that 1.5 g of root tissue is lost for every g of chlorotic leaf dry weight.

Table 28. Carrot responses to O_3.

O_3[y] (pphm)	Leaf length (cm)	Root length (cm)	Total no. leaves	% chlorotic leaves	Total fresh wt (g)	Total dry wt (g)	Root dry wt (g)	Leaf dry wt (g)	Root/ shoot ratio
0	45.7a[z]	21.9a	34.1a	2a	351a	44.5a	24.9a	19.6a	1.23a
19	50.2ab	19.9a	39.2a	14b	272b	39.1a	16.9b	22.2b	.73b
25	52.4b	20.1a	39.8a	28c	244b	32.6b	13.4b	20.0a	.60b

[y] Intermittent 6 hr exposures during about 6% of the total growth period.
[z] Mean separation in columns by Duncan's multiple range test, 5% level.
Data from Bennett and Oshima (1976)

The problem of insecticide residues in carrots sprayed to control rust fly maggot has been studies using different rates and methods of application of four insecticides (Finlayson 1976). The concentration of residues within the carrot is greater in the peel than in the pulp and greater in the top portion of the carrot than in the lower sections. Residues increase in proportion to rates and numbers of applications.

8.5 Celery

There has been very little characterization of celery responses to pollutants. Comparable O_3 injury occurs in controlled exposures or in the field (Proctor and Ormrod 1977). However, cultivars differ in relative sensitivity between the two environments.

8.6 Corn

This species is rated as sensitive to O_3, N, F and Cl_2, and insensitive to SO_2 and PAN (Jacobson and Hill 1970). Sweet corn and field corn responses to pollutants have had relatively little study in relation to the value of the crop, possibly because the large size of mature plants makes it comparatively difficult to conduct whole plant exposure experiments.

Oxidants. Ozone injury is characterized by bands of interveinal bifacial and upper surface necrosis (Fig. 44). As for other species, widely differing

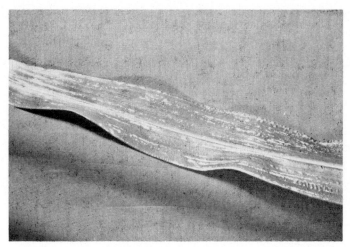

Fig. 44. Ozone injury on corn leaf blade as a result of controlled exposure (Author's photo).

sensitivity of sweet corn cultivars to oxidant smog can be noted in the field in oxidant-prone areas such as California, U.S.A. (Cameron et al. 1970). Controlled exposures to O_3 at concentrations lower than those in air surrounding urban areas result in significant yield reductions in the cultivar 'Golden Midget' but not in 'White Midget' (Heagle et al. 1972). Exposure of sweet corn to O_3 in glasshouses results in significant reductions in yield and overall loss of vigour and stature (Oshima 1973). Injury comparable to that caused by controlled O_3 treatments occurs on plants exposed to oxidant smog in the field (Cameron and Taylor 1973) suggesting that it is the O_3 component of oxidant smog which is causing the injury.

Filtered and non-filtered outdoor chambers have been used to determine the relative susceptibility of two sweet corn hybrid cultivars to ambient photochemical oxidants (Thompson et al. 1976). Plant height, tiller number and length, and stalk weight are less in non-filtered air. Cultivars differ in sensitivity, with the number and weight of marketable cobs and number of fully developed kernels severely reduced by oxidants. Reductions in fruit number and weight in a sensitive cultivar are accompanied by some leaf tissue collapse and premature yellowing of leaves. Similar differences in foliar symptoms between cultivars can be noted in open field plantings in an oxidant-pollution area.

Other Air Pollutants. There has been interest in the response of corn to HF (Leone et al. 1956). Few other studies have involved corn as an experimental species, with the result that the interactions of air pollutant response with environmental and plant factors are virtually unknown.

Heavy Metals. Corn has been a favoured crop for studies of the disposal of sewage sludge on land, because the management systems for this crop might easily incorporate a sludge disposal program. This interest has led to numerous controlled studies of corn responses to heavy metals and/or sewage sludges. An optimal rate of sewage sludge application can be established with which yield will be stimulated by the addition of N, P and K nutrients (Cunningham et al. 1975a). However, this rate will result in an increase in tissue heavy metal contents. Toxicity, particularly due to Cu, occurs at supraoptimal sludge rates. Amendment of sludge with Cu, Cr, Ni and Zn salts has been undertaken to determine effects on corn yield and metal uptake from soil limed to pH 6.8 (Cunningham et al. 1976b). Yields decrease as the Cu and Zn concentration in the sludge increase, Ni has no appreciable effect, and yield increases with Cr, indicating suppression by Cr of the uptake or distribution of other metals. Corn yield reductions are less after an intervening rye crop.

The Pb content of corn has also been of interest in relation to possible Pb pollution of corn-growing soils near highways or industries. No Pb toxicity is observed when excessive amounts of Pb are applied to field soil on which corn is grown (Baumhardt and Welch 1972). Up to 3200 kg/ha Pb may be added with no detrimental effect on growth or kernel yield. While the Pb content of vegetative parts of the plant is affected by the Pb treatments, the kernels contain only 0.4 ppm Pb and are not affected by the Pb added to the soil. Corn root responses to Pb and Cd have been studied (Hassett et al. 1976).

8.7 Cress

Cress plants have been used to demonstrate that above-normal amounts of heavy metals in plant tissue may increase O_3-induced leaf injury. Application of Cd or Zn to cress plants results in markedly increased O_3-induced phytotoxicity in terms of visible leaf injury and pigment degradation (Czuba and Ormrod 1974).

8.8 Cucumber

Cucumber leaves are sensitive to O_3 with wide variations in amount of leaf chlorotic mottle among cultivars (Ormrod et al. 1971) (Fig. 45). Physiological responses of cucumber to O_3 have been characterized (Frick and Cherry 1974). Chlorophyll reductions in actively growing leaves are observed and amino acid, nucleic acid and protein reactions are very dependent on elapsed time after O_3

Fig. 45.   Apparent O_3 injury on cucumber leaves resulting from ambient O_3
exposure (Author's photo).

treatment.   Cucumber leaves have been used to study air flow rate effects on
severity of injury (Brennan and Leone 1968).   The severity of either O_3 or
SO_2 injury is greater at higher flow rates.

Cucumber seedlings can be protected from O_3 injury by means of prolonged
contact with a wide variety of antioxidant chemicals in solution, including
hydrazine, indole, tryptophane and mescaline (Siegel 1962).   Ascorbic acid is
ineffective in this species.   Protection is evident in increased survival of
seedling populations and in reduced inhibition of hypocotyl elongation.   Compounds with little or no antioxidant activity fail to protect against O_3 injury.

8.9   Eggplant

There is a wide range of eggplant cultivar sensitivity to O_3 (Rajput and
Ormrod 1976).   Ozone injury ranges from necrotic flecking to bleaching of
interveinal areas (Fig. 46).

8.10   Lettuce

This species is relatively sensitive to SO_2, NO_2 and PAN (Jacobson and
Hill 1970).   Cultivar sensitivity to O_3 has been evaluated, with 'Dark Green
Boston' the most sensitive of eight cultivars tested and 'Great Lakes' and
'Black-seeded Simpson' the least sensitive (Table 29).

Fig. 46. Ozone injury to 'Imperial Black Beauty' eggplant leaf of intermediate
age exposed to 40 pphm O_3 for 6 hr each day on 2 consecutive days
(Rajput and Ormrod 1976).

Table 29. Ozone sensitivity of lettuce cultivars.

Cultivar	Plant injury	Cultivar	Plant injury
Dark Green Boston	45.7a[z]	Big Boston	23.6c
Grand Rapids Forcing	30.3b	Romaine	19.3d
Imperial #456	24.1c	Simpson, Black Seeded	11.2e
Butter Crunch	23.7c	Great Lakes	9.9e

[z] Average % injury of the whole lettuce plant after exposure to 70 pphm O_3
for 1.5 hr. Mean separation by Duncan's multiple range test, 5% level.
Data from Reinert et al. (1972b).

     Nitrogen nutrition affects response to oxidants. Application of N to the
soil at the rate of 45 kg/ha increases the oxidant injury to romaine lettuce
by 80%, compared to control plants receiving no N additions. Lower temperatures
decrease sensitivity (Middleton 1956). Increasing the number of days of $13^{\circ}C$
exposure decreases the sensitivity of romaine lettuce to oxidant compared with
plants exposed to the same concentration of oxidants but maintained at $24^{\circ}C$.

Lettuce responses to HF have been described based on continuous exposure of plants for up to 4 months (Benedict et al. 1964). Top growth is unaffected unless there are visible injury symptoms on the leaves. Fluoride translocates to the leaf margins with no downward translocation.

Lettuce has also been a test species for heavy metal studies. Lettuce plants readily take up Cd from Cd-treated soils and may easily accumulate potentially hazardous concentrations (John et al. 1972). Plants grown in nutrient solution containing Cd have been used to determine the relationship of Cd to other elements in the nutrient solution and in the plant (John 1976). Increases in solution K or pH reduce Cd uptake, and Cd levels affect tissue P, Fe, Mn, Al and Ca concentration. Lettuce cultivars do not differ greatly in visible Cd toxicity but there are cultivar differences in dry matter yields and concentration in the leaves (John and Van Laerhoven 1976a). Studies of responses of soil-grown lettuce to Cd indicate that addition of Cd to soils usually decreases yield. Higher soil organic matter decreases tissue Cd, as does liming of some acid soils. Addition of Cd increases tissue Cd concentration but added Zn does not affect Cd uptake.

8.11 Muskmelon

Oxidant sensitivity is a noteworthy characteristic of this species with a diffuse chlorotic mottle developing first on older leaves (Jacobson and Hill 1970). Diphenylamine dusts protect muskmelons from O_3 injury (Gilbert et al. 1975).

8.12 Onion

Sensitivity to O_3 and Cl_2, and insensitivity to PAN and ethylene character- ize this species (Jacobson and Hill 1970). Research projects have been rela- tively limited, with the predominant studies involving oxidants, including a classic study of the mode of action and inheritance of O_3 resistance (see Plant Breeding and Genetics). There is a relationship of onion leaf tipburn to O_3 injury (Engle et al. 1965). Leaf flecking occurs first as a result of O_3 expo- sure and rain bruising during storms. Leaf tissues are destroyed or altered, resulting in impaired translocation of metabolites to the leaf tip. The leaf tip thus dies back (Fig. 47).

The interaction of O_3 injury on onions with _Botrytis_ spp. has been studied by exposing plants to antiozonants and/or fungicides (Wukasch and Hofstra 1977a). Fungicide-antiozonant combination treatments are most effective in reducing _Botrytis_ infection. Open-top chambers have been used to study onion leaf die- back and yield reduction in an O_3-prone area (Wukasch and Hofstra 1977b). Removal of 35 to 65% of the ambient O_3 by carbon filters results in marked

130

Fig. 47. Onion leaf tipburn associated with ambient O_3 episodes (Photo
courtesy of G. Hofstra).

reduction of leaf dieback and a 28% increase in yield. Inoculation of wetted
leaves with suspensions of <u>Botrytis</u> <u>squamosa</u> results in more lesions and dieback
in an unfiltered chamber suggesting that higher O_3 results in greater Botrytis
sensitivity in onions.

Dithane fungicides cause abnormalities in onion chromosome behavior with
resultant reduced seed set (Mann 1977).

8.13 <u>Peas</u>

Biochemists have long favoured peas for studies of metabolic responses to
pollutants but there have been very few studies of whole plant responses. The
species is rated as sensitive to SO_2 and ethylene (Jacobson and Hill 1970) and
also to O_3 (Ormrod 1976) (Fig. 48).

<u>Oxidants</u>. Exposure of 'Alaska' peas to ozonated gasoline or ozonated
hexene results in temporary slowing of shoot elongation (Todd and Garber 1958).
The treated plants recover and grow normally one or two days later. Elongation
of this cultivar is not slowed by O_3 or 1-hexene alone. Several processing pea

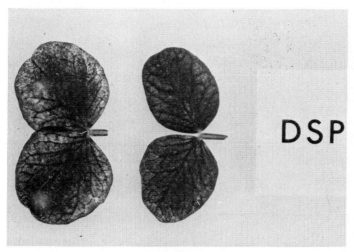

Fig. 48.   Injury symptoms on the upper surface of 'Dark Skin Perfection' pea
leaves exposed to O_3 (Ormrod 1976).

cultivars have been evaluated for O_3 sensitivity (Ormrod 1976).   They vary in
sensitivity to 25 pphm O_3 but are similarly injured by 50 pphm O_3.   Some cultivars also differ widely in individual plant responses, suggesting variation in
genetic tolerance to O_3.   The heavy metal enhancement of O_3 toxicity has been
demonstrated for peas (Ormrod 1977).   Addition of one or 10 μM $CdSO_4$ or 10 or
100 μM $NiSO_4$ to nutrient solutions followed by O_3 exposure results in more
injury than from O_3 alone (Fig. 49).   Higher metal levels suppress both plant
growth and O_3 toxicity.   Responses to trace elements and O_3 vary somewhat with
cultivar used.   Peas exposed to NO_2 have concurrent leaf necrosis and increased
protein content (Zeevaart 1976).

Sulphur Dioxide.   Pea plants have been used to study methods of diagnosing
SO_2 injury before visual symptoms occur (Jager and Klein 1977).   Exposure to
concentrations of 10 or 15 pphm causes accumulation of inorganic S, reduced
buffer capacity of cells, and stimulation of glutamate dehydrogenase activity,
along with slight leaf chlorosis.   Plants exposed to 25 pphm SO_2 have considerable foliar necrosis and weight reduction in addition to the effects found at
the lower concentrations.   The three biochemical changes are thus suggested as
early indicators of SO_2 injury.

Fig. 49. Ozone injury enhancement in pea leaves by Ni. C - control, Ni - Ni only, O_3 - O_3 only, Ni + O_3 - combined treatment. Ni - 100 µM $NiSO_4$ in nutrient solution, O_3 - 50 pphm for 4 hr (Ormrod 1977).

8.14 Potato

Some potato cultivars are notoriously sensitive to oxidant pollutants but little research has been conducted on the response of this important horticultural species to other pollutants.

Oxidants. The characteristic dark flecking symptom on potato leaves exposed to oxidant has resulted in oxidant injury being commonly called "speckle leaf" disorder. The wide range of oxidant sensitivity of potato cultivars has been demonstrated by several investigators. However, there is not complete agreement among studies, suggesting that cultivar environment interactions are an important aspect of oxidant responses of potatoes.

In Michigan, U.S.A., "speckle leaf" of potatoes had been noticed for years without associating it with air pollution injury (Hooker et al. 1973). Sensitive cultivars have lower yields associated with early vine maturity, premature vine death and yellowing of lower leaves. Affected leaves first develop necrotic spots on the upper surface. This is followed by bronzing and upward leaf rolling (Fig. 50). Severely injured leaves become chlorotic and

Fig. 50.   Ozone injury on potato leaves characterized by necrotic spots, bronzing, and upward leaf rolling (Photo courtesy of G. Hofstra).

remain attached to the stem. Field studies in Delaware, U.S.A., indicate a wide range of sensitivity to oxidant smog among potato cultivars and breeding strains (Brasher et al. 1973). Yield losses of up to 50% are estimated for some sensitive cultivars in Atlantic coastal U.S.A. (Heggestad 1973; Leone and Green 1974).

A wide range of O_3 resistance of potatoes has also been demonstrated by exposure of plants to controlled O_3 treatment (Brennan et al. 1964). Some cultivars have leaf injury after four hr at 10 pphm O_3, others are intermediate, and one cultivar is uninjured at 56 pphm O_3 (Table 30).

Environment effects. As for other species and pollutants there are likely to be many interactions involved in the response of potatoes to oxidant smog. For example, the "speckle leaf" disorder can be associated with high soil

Table 30. Ozone injury to potato leaves.[y]

Cultivar	O_3 concn pphm				
	5	10	23	35	56
Chippewa	0[z]	X	XX	XX	XX
Cobbler	0	X	XX	XX	XX
Plymouth	0	X	XX	XX	XX
Pungo	0	X	XX	XX	XX
Katahdin	0	0	X	X	XX
Green Mountain	0	0	0	X	XX
Kennebec	0	0	0	X	XX
Avon	0	0	0	0	0

[y] Exposed to 4 hr O_3.

[z] Leaf injury ratings: 0 = no injury, X = slight injury, XX = severe injury.

Data from Brennan et al. (1964)

acidity, and it is suggested that the resultant nutritional imbalance may affect plant response to oxidant (McKeen et al. 1973). Discrepancies between the cultivar ratings by various investigators are apparent when all presently available data are tabulated. This suggests that differences in physiological status of the plants and environmental conditions at time of exposure are sufficient to change the sensitivity relationship of cultivars.

"Speckle leaf" appeared last in the highest N treatment in field experiments (Vitosh and Chase 1973). Petiole analyses revealed that the stage of maximum N uptake by the potato plant corresponds very closely with the initiation and development of "speckle leaf". Leaves deficient in N at this stage are the most susceptible to oxidant injury. The disorder is also enhanced by additional irrigation (Table 31).

The interaction of O_3 and plant pathogens has been illustrated by experiments with potatoes (Manning et al. 1969). Ozone injury in controlled exposure conditions apparently increases potato leaf susceptibility to Botrytis cinerea, corroborating field observations.

Other Air Pollutants. Exposure of potato plants to ethylene results in the formation of large numbers of small tubers suggesting a role for ethylene in tuberization (Posthumus, unpublished). An unidentified form of air pollution injury on potatoes has also been described (Kirkham and Keeney 1974). Distinct injury symptoms, not resembling those due to any major known pollutant, can be noted on potato plants in growth chambers.

Table 31. Potato yield and "speckle leaf" incidence on 'Haig' and 'Norchip'
grown with different N levels and water management.

Sidedress N (kg/ha)	Additional water		Normal water	
	Haig	Norchip	Haig	Norchip
0	130^z severe	230 moderately severe	307 slight	353 slight
84	172 severe	256 moderate	362 slight	354 slight
168	347 slight	353 slight	354 slight	411 slight

^{z} Yields in q/ha.

Data from Vitosh and Chase (1973)

Trace Elements. The widespread use of metal-containing insecticides in
potato production over the last several decades has led to several studies of
metal residues. For example, the use of As compounds to control insects and
kill vines results in the accumulation of As in soils (Sinclair et al. 1975).

Accumulation of Hg in potato foliage is proportional to the number of
sprays of MCO (Ross and Stewart 1964). Translocation of Hg from foliage to
tubers is evident with Hg accumulation in the tubers during the period of
rapid increase in size (Table 32).

Table 32. Mercury residues (ppm) in potato foliage and tubers from different
spray schedules of MCO.

Sprays applied	Foliage	Peel	Pulp	Whole Tuber
1,2	1.1	0.064abcd^z	0.038abc	0.042ab
1-4	3.0	0.093abc	0.037abc	0.039abc
1-6	5.4	0.089abc	0.038abc	0.046ab
1-8	6.8	0.118a	0.056a	0.065a
3-8	6.3	0.100ab	0.044ab	0.041ab
3-6	5.8	0.058bcd	0.029abc	0.034bc
5-8	5.3	0.082abc	0.037abc	0.042ab
7,8	2.9	0.039cd	0.017bc	0.020bc
0	0.5	0.026d	0.011c	0.013c

^{z} Mean separation by Duncan's multiple range test, 1% for peel and 5%
level for pulp and whole tuber.

Data from Ross and Stewart (1964)

Pesticides. A potato crop on a sandy loam soil was used in the development of a computer model to estimate leaching of pesticides (Leistra and Dekkers 1976). The uptake by the crop is an important factor in reducing leaching, along with decomposition, particularly for weakly absorbed and comparatively persistent compounds.

8.15 Radish

Long a favourite species for pollution research, radish is rated as sensitive to O_3, SO_2, Cl_2 and H_2S, but insensitive to PAN and ethylene (Jacobson and Hill 1970). The species is widely used for both air and soil pollution studies. Its rapid growth and sensitive root/shoot relationships make radish an ideal species for study.

Oxidants. Radish plants exposed to O_3 have more reduction of root growth than of leaf growth (Tingey et al. 1971a, 1973a). Also, root growth reduction is much greater than would be expected based on foliar injury symptoms. Plants are most sensitive in terms of root growth reduction just prior to the stage of rapid root enlargement. The magnitude of O_3-induced root growth reductions depends on frequency of exposure as well as development stage. Cultivar sensitivity evaluations indicate that 'Cherry Belle' is the most sensitive of nine radish cultivars tested, and 'Icicle' the least sensitive (Table 33).

Table 33. Ozone sensitivity of radish cultivars.

Cultivar	Plant injury[z]
Cherry Belle	34.7a
Crimson Giant	33.9a
Comet	32.4ab
Champion	30.7abc
Red Boy	24.7bcd
Calvalrondo	23.7cd
Early Scarlet Globe	23.6cd
French Breakfast	23.4cd
Icicle	17.1d

[z] Mean % injury of the 3 most severely injured leaves after exposure to 35 pphm O_3 for 1.5 hr. Mean separation by Duncan's Multiple Range Test, 5% level.

Data from Reinert et al. (1972b)

The interactions of environment with O_3 response have also been studied in radish. In temperature studies, cultivars react differently to O_3 in response to pre- and post-exposure temperatures (Adedipe and Ormrod 1974). In 'Cavalier', O_3 decreases leaf weight of plants grown at a pre-exposure temperature of $20/15^{\circ}C$ day/night, but not of those grown at $20/25^{\circ}C$. Post-exposure temperature has no effect on O_3 symptoms. In 'Cherry Belle', O_3 decreases leaf weight only of plants held at $20/15^{\circ}C$ during the post-fumigation period. Responses of radish to O_3 when grown at different levels of N or P are modified by the temperature regime (Ormrod et al. 1973). Growth suppression due to O_3 is similar between N levels at both $20/15$ and $30/25^{\circ}C$ day/night temperatures while growth suppression in either high or low P occurs at $20/15$ but not at $30/25^{\circ}C$.

An evaluation of the use of growth regulators to protect radish plants against O_3 indicated that O_3 has no effect on leaf weights of 'Cavalier' radish plants pretreated with 30 mg/l BA, GA or IAA, but decreases leaf weight of control plants and of plants pretreated with ethephon (Adedipe and Ormrod 1972). However, O_3 decreases the root weight of all plants except those treated with BA. Benzyladenine is the most active of the growth regulators tested for the protection of radish plants from O_3 injury, in preventing growth suppression of both root and shoot and in preventing a decrease of leaf chlorophyll content.

Other Air Pollutants. Exposure of radish seeds to HCN, H_2S, NH_3, Cl_2 and SO_2, at 250 or 1000 ppm for up to 960 minutes showed that germination is delayed by HCN and H_2S but % germination is not affected, while NH_3 and Cl_2 are more toxic, and SO_2 is most toxic (Barton 1940). Moist seeds are more sensitive than dry.

Radish exposure to mixtures of O_3 and SO_2 results in some growth parameters being affected additively and others less than additively (Tingey et al. 1971a; Tingey and Reinert 1975). Ozone-SO_2 interactions may result in visible injury enhancement by the mixture compared to the effects of the separate gases (Fig. 51).

Trace Elements. Radishes are often grown in locations exposed to metal contamination, and the response of this species to some of the trace elements has been characterized. Air and soil, but not water, are the significant sources of tissue Pb in radishes (Dedolph et al. 1970; Ter Haar et al. 1969). However, Pb in the edible root is unaffected by variations in soil or air Pb. Radish responds to $CdCl_2$ added to the soil, and shoot and root Cd concentrations increase with increasing Cd addition (John 1972b). Shoots accumulate more Cd than roots. Adding lime to the soil decreases Cd uptake (Table 10).

138

Fig. 51.   Radish leaf responses to O_3 and/or SO_2 (Rajput et al. 1977).

Soil additions of Cd, Pb or Zn result in relatively small changes in radish plant composition in response to large applications (Lagerwerff 1971). Soil pH increase decreases metal uptake. Aerial contamination is estimated to account for 40% of the metal content when plants are grown 200 m from a busy roadway.

8.16 Spinach

Spinach is sensitive to both O_3 and SO_2 with intermediate sensitivity to PAN (Jacobson and Hill 1970). Its sensitivity makes it suitable for use as an indicator of air pollution episodes during cooler parts of the growing season. The occurrence of O_3 injury to several vegetables, particularly spinach, has been reported in New Jersey, U.S.A. (Daines et al. 1960b). The upper leaf surface has typical necrotic injury and the tissues of affected leaves become necrotic and sunken. The O_3 sensitivity of spinach cultivars has been evaluated (Manning et al. 1972b). There is a wide range of sensitivity with the most sensitive cultivars injured after four hr of 10 pphm O_3 and the greatest range of response at 15 and 20 pphm. New Zealand spinach is less sensitive than standard cultivars.

Three early studies focused on the effect of N fertility on oxidant res- ponse. The incidence and severity of injury resulting from ozonated hexene exposure is significantly increased as the N level increases (Brewer et al.

1962). Exposure of spinach plants to peroxides derived from olefins results in five to seven times as much injury to plants grown with abundant N supply as to those grown under low or deficient N status (Kendrick et al. 1953). The application of N to soil at the rate of 45 kg/ha increases the amount of oxidant injury to spinach leaves by 40%, compared to plants not receiving N additions (Middleton 1956).

Other nutrient relationships have also been studied (Kendrick et al. 1953). For P the relationships are much more complex than for N, with interactions of P with N and with K. Phosphorus additions alone decrease oxidant injury; K without P usually increases injury at low but not at high N. Addition of both K and P decreases injury severity. Lower temperatures result in less oxidant sensitivity.

Spinach plants have been exposed continuously to HF for up to four months to observe the effects of this pollutant (Benedict et al. 1964). Top growth is not affected unless there are visible markings on the leaves, while root growth is slightly reduced by any HF exposure. Fluoride is translocated to the leaf extremities, with no downward translocation.

Physiological responses of spinach to H_2S have been described (Oliva and Steubing 1976). A decline in photosynthesis is measured in an H_2S concentration of 1.2 mg/m^3. Water loss and leaf necrotic injury symptoms are observed at 2.5 mg/m^3. Respiration is less sensitive to H_2S than is photosynthesis.

8.17 Tomato

Tomato has been the subject of many air pollution oriented projects even though visible injury is seldom seen in the field. The species is rated as relatively sensitive to O_3, PAN, ethylene, 2,4-D, HCl and H_2S but insensitive to F (Jacobson and Hill 1970).

Oxidants. The fresh weight of tomato tops exposed to ozonated hexene, a mixture of oxidants, during the lighted hours is consistently less than the fresh weight of comparable control plants (Koritz and Went 1953). There is no effect on growth when plants are exposed in the dark, in light just following a 12-hr dark period, or in the dark after 1 1/2 hr of light, thus providing indirect evidence of the relationship of pollutant sensitivity to stomatal opening. The size of tomato leaves is reduced on plants grown in polluted ambient air in California, U.S.A., but the numbers of nodes and inflorescences are not different from those of plants grown in carbon-filtered air (Taylor 1958). However, 60% of the blossoms abort in oxidant-containing air, while almost all blossoms on plants grown in filtered air set fruit.

Open-top field chambers were used to demonstrate that yields of a commercial tomato cultivar are reduced at least 33% by ambient oxidant at Yonkers, New York, U.S.A. (MacLean and Schneider 1976). This suggests that oxidant sensitivity may be an important economic problem in tomato production even though, unlike potatoes and some other vegetables, severe visible injury symptoms have not been noted in the field.

Ozone. The symptoms of O_3 injury on tomato have been described in detail and range from chlorotic stippling on the upper surface of leaves to monofacial or bifacial necrotic areas (Gentile et al. 1971). Severe cases are characterized by general chlorosis or premature senescence along with abscission of leaves (Fig. 52). The mature leaves are sensitive to O_3 while young leaves

Fig. 52.   Ozone injury on leaves from six cultivars of tomato ranging from insensitive to very sensitive (Photo by Neil and Ormrod, unpublished).

are relatively insensitive. Shoot length is increased by O_3 exposure regardless of the visible injury (Neil et al. 1973). Of several Lycopersicon species examined for O_3 sensitivity, L. esculentum is least sensitive and L. pimpinellifolium most sensitive of seven species tested (Table 34). Pollen germination is reduced by 40% in the sensitive L. pimpinellifolium. Insensitivity to air pollutants does not always coincide with resistance to diseases (Ten Houten 1974). For example, L. pimpinellifolium is successfully used in breeding programmes as a source of disease resistance, but it has high O_3 sensitivity.

Table 34. <u>Lycopersicon</u> species evaluated for sensitivity to O_3.

Sensitivity classification	Species name	No of introductions
Insensitive	L. esculentum	5
Intermediate	L. esculentum	8
	L. glandulosum	2
	L. hirsutum f. glabratum	3
	L. peruvianum	7
Sensitive	L. esculentum	9
	L. glandulosum	1
	L. pimpinellifolium	5
	L. hirsutum	3
	L. hirsutum f. glabratum	3
	L. peruvianum	8
	L. peruvianum var. humifusum	2

Data from Gentile et al. (1971)

More than 1200 lines in a world tomato collection have been tested for O_3 sensitivity and only four were insensitive (Clayberg 1971). Two additional insensitive cultivars are found in a North American group. Marked differences in sensitivity are also found among commercial cultivars (Reinert et al. 1972a). Tomato yield responses in the field seem to be related to oxidant smog conditions early in the flowering season and F_1 hybrids are superior to non-hybrid cultivars under conditions of severe atmospheric pollution (Lesley and Taylor 1973). Exposure of tomato plants to O_3 periodically for 15 weeks results in extensive foliar injury, defoliation and reduced plant growth, but yield itself is less affected than are other parameters (Oshima et al. 1975). Fruit yield may not be reduced even with a 27% reduction in total plant dry weight, indicating a higher threshold in tomato for yield effects than for vegetative growth effects (Table 35).

Exposure of 'Tiny Tim' tomatoes to O_3 in the glasshouse repeatedly for two to three months results in reduction in root and shoot growth, plant development and fruit yields (Manning and Feder 1976). There is interference with fruit set and premature senescence of immature fruit. Fresh market tomato cultivars planted at a high and a low O_3 location in California, U.S.A. were used to show that O_3-induced foliar injury is not correlated with yield response (Oshima et al. 1977a). Plants at the high O_3 location have reduced fruit size and depressed early season production with significant differences among cultivars. At the low O_3 location most cultivars yield similarly. The lack of association of foliar injury with yield reductions has important implications because much of the data available on species and cultivar sensitivities to O_3 is for foliar injury, rather than for yield.

Table 35. Ozone effects on 'H-11' tomato plants and harvested fruits.

Plant response	O_3 concentration (pphm)[y]		
	0	20	35
Fresh weights (g)			
Leaves and stems	755 a[z]	549 b	285 c
Immature fruit	260 a	295 a	104 b
Dry weights (g)			
Leaves and stems	167 a	114 b	46 c
Immature fruit	17.3 a	17.6 a	7.3 b
Roots	14.3 a	12.7 a	5.9 b
Number			
Immature fruit	12.8 a	11.7 a	3.9 b
Flowers	6.2 a	10.7 a	9.8 a
Harvested fruit			
Fruit height (cm)	4.6 a	4.9 a	4.5 a
Fruit diameter (cm)	5.5 a	5.7 a	5.3 a
Average fruit weight (g)	76 a	86 a	68 a
Number of fruit	11.3 a	9.9 a	6.9 b

[y] 2.5 hr/day, 3 days/wk for 15 wks.

[z] Mean separation in rows by Duncan's multiple range test, 1% level.

Data from Oshima et al. (1975). Reproduced from JOURNAL OF ENVIRONMENTAL QUALITY, 4, 463-4, by permission of the American Society of Agronomy, Crop Science Society of America and Soil Science Society of America.

It was confirmed in another study that O_3 concentrations under outdoor conditions have important effects on tomato yield in California, U.S.A. (Oshima et al. 1977b). Eleven locations, encompassing a gradient of ambient O_3 concentrations, were used to determine that 85% of the reduction in fruit size is due to O_3 dose and that O_3 concentration is much more important than any other meteorological variable. A cumulative dose of 2000 pphm-hr greater than 10 pphm is predicted to reduce yield by 50%.

Age of tomato leaves has an important effect on their oxidant sensitivity. Exposure of tomato plants to controlled oxidant levels results in plant injury indices for successive leaves from base to apex of 0, 0, 1, 3, 2.9, 6.5, 9.4, 5.2, 1.6 and 0 (Middleton et al. 1958).

Environment effects. Early studies of environment effects on tomato response to synthetic smog (1-n-hexene plus O_3) showed that mid-day exposure is followed by decreased growth without visual injury while exposure in the dark or at the beginning of the light period has no effect (Koritz and Went 1953). Provision of ample water increases susceptibility and transpiration rates decline upon smog exposure. When plants are subjected to repeated exposures, there are repeated decreases in transpiration followed by partial recovery. The observation that time of exposure is important has been corroborated, with the plants more sensitive to O_3 in the afternoon than in the morning (Reinert et al. 1972a). Air flow rate is another consideration. Tomato leaves have greater injury due to either O_3 or SO_2 at a high than at a low flow rate (Brennan and Leone 1968).

Studies of soil and plant water status effects indicate that O_3 decreases dry matter production of optimally-watered tomato plants, whereas plants subjected to moderate water stress resulting in low leaf turgidity at exposure are protected from O_3 toxicity (Khatamian et al. 1973). Stomatal opening is affected by soil water stress and determines O_3 toxicity (Adedipe et al. 1973).

Soil O_2 diffusion rates are related to tomato plant susceptibility to O_3 in the glasshouse (Stolzy et al. 1961). Low O_2 diffusion rates make the plants less O_3 sensitive, and this effect is most noticeable when plant vigour is obviously impaired. The effect of reduced O_2 in the root zone on tomato response to O_3 can be studied by means of ventilated root chambers (Stolzy et al. 1964). Reduced rates of soil O_2 diffusion reduce the susceptibility of the plants to O_3 injury.

Toxicity of O_3 in tomato increases with increase in P supply (Leone and Brennan 1970). Visible injury due to O_3 is reduced by K deficiency (Leone 1976). Foliage diffusion resistance is greater in K-deficient than in normal plants. Stomatal opening is considered to be related to the accumulation of K in guard cells and foliar K content increases after O_3 exposure in both normal and K-deficient plants.

Protection. Tomato foliage is protected from atmospheric O_3 injury by the application of chemicals to overhead shade cloth (Rich and Taylor 1960). Manganous 1,2-naphthoquinone-2-oxime provides protection as do the similar cobaltous and manganous chelates of 8-quinolinol. Antioxidants used in the rubber industry, such as dialkyl-p-phenylene diamines, are even more effective. Tomato is protected against O_3 when grown in soil containing 9.5 ppm carboxin (Rich et al. 1974) or when sprayed with EDU (DPX4891) (Fig. 53). Protection of tomato plants against O_3 is obtained by the use of darkness, soil-plant water stress, and PMA to close stomata (Adedipe et al. 1973).

Interactions. The interaction of selective herbicides with O_3 exposure of tomato has been studied (Carney et al. 1973). Herbicide and O_3 toxicity are additive with no interaction when diphenamid is applied to tomato followed by O_3 exposure. In contrast, the two have synergistic effects in tobacco. Application of diphenamid to tomato followed by exposure of the plants to O_3

144

Fig. 53.   Chemical protection of tomato from O_3 injury. Control (C), benomyl
sprayed (B) and EDU (DPX4891 – an experimental Dupont chemical)
sprayed (DPX) (Author's photo).

demonstrated that the time course of diphenamid metabolism is altered by O_3
exposure allowing elucidation of some aspects of diphenamid metabolism (Hodgson
et al. 1974). Tomato plants survive otherwise injurious dosages of 2,4-D and
O_3 if the two are applied simultaneously (Sherwood and Rolph 1970). Zinnia and
elm plants respond similarly to tomato.

Peroxyacetyl nitrate. Widespread PAN-type injury on recently transplanted
field tomatoes was found in Ontario, Canada (Pearson et al. 1974). Injury
symptoms range from under-surface silvering, glazing and bronzing, to bifacial
necrosis, and are associated with weather conditions favourable to the formation
and transport of PAN-type oxidants.

Nitrogen dioxide. Responses of tomato to NO_2 were studied by exposing
tomato seedlings continuously to low concentrations for 10 to 22 days (Taylor
and Eaton 1966). There is significant growth suppression, darker green colouration with increased chlorophyll content, and downward curving of leaves.
Spierings (1971) also studied NO_2 effects on tomato, exposing plants for 21 to
45 days and finding increased shoot length but decreased leaf size and petiole
length. Exposure of plants to 25 pphm NO_2 for the entire growth period results
in the development of smaller, duller green leaves with shorter petioles, and
decreases fruit yield by 22%. Growth is generally retarded and senescence and

abscission are hastened. Glasshouse tomatoes are more sensitive to NO_x than are several other species (Capron and Mansfield 1976).

Sulphur Dioxide. Tomato plants deficient in N or S exhibit decreased susceptibility to SO_2 (Leone and Brennan 1972a, b), while supra-optimal S increases and N decreases susceptibility compared with optimum nutrient levels.

Fluoride. Toxicity symptoms on tomato due to F are characterized by tip and marginal scorching of leaves whether F is applied through the roots or by exposure of shoots (Brennan et al. 1950; Leone et al. 1956). Severity of leaf injury is directly correlated with F exposure concentration. Fluoride injury of tomatoes has been investigated in an attempt to quantify "hidden injury"; the occurrence of growth reductions without visible injury symptoms (Hill et al. 1957). Photosynthesis rates are not reduced, plant height is unaffected and chlorophyll content unchanged at F concentrations up to that which causes visible visible leaf injury, suggesting that F-induced hidden injury does not occur in tomatoes. Some metabolic effects of F on tomato leaves also have been reported (Weinstein 1961).

Tomato growth and fruiting are not affected by a 99 day exposure to 0.6 mg F/m^3 during the growth period (MacLean et al. 1977). Germination of pollen from tomato plants after long-term continuous HF exposure has been evaluated (Sulzbach and Pack 1972). Reduced germination is found only if the plants are also grown at low Ca levels. Cucumber pollen germination is not similarly affected.

Nutritional status affects F toxicity in tomato (Brennan et al. 1950). Deficiency of N, Ca or P reduces the uptake of F from soil or air. Excess N or Ca also decreases F toxicity. The top and margin necrosis characteristic of F toxicity increases with increase in P supply, but is minimal at both deficient and excessive levels of N. Both deficient and excessive amounts of Ca also reduce F injury. The effect of Ca concentrations above 40 ppm is to precipitate F in the form of insoluble compounds which reduce the uptake of F and thus the foliage injury. Exposure of tomatoes of differing Ca nutrition status to six mg F/m^3 throughout flowering and fruit development causes fruit size reduction (Pack 1966). Foliage is injured by lower HF concentrations. The nature of the F injury suggests that F interferes with Ca metabolism.

Calcium, Mg and K deficiencies also affect the HF responses of tomato (MacLean et al. 1969). Exposure to HF increases the apparent severity of Mg deficiency. Plants deficient in Ca are also more stunted and chlorotic in the presence of HF. Deficiency of K has least interaction with HF. Foliar accumulation of F is decreased by Mg deficiency, unaffected by Ca deficiency and enhanced by K deficiency. Further studies of Mg nutrition effects on F susceptibility of tomato have been reported (MacLean et al. 1976). Chlorosis induced by F is greater in Mg-deficient plants and less in high Mg plants compared to plants grown at the Mg level of a complete nutrient solution.

Other Air Pollutants. Field tomato production on the island of Hawaii, U.S.A. can be severely affected by volcanic air pollution (Kratky et al. 1974). A mixture of phytotoxic compounds is apparently carried on to the plants by rainwater and results in blossom drop and poor set of hollow, small, almost seedless fruits. Pollen germination and pollen tube growth are inhibited. Plants under plastic rain shelters grow normally. The first portion of a rainfall is most injurious.

Trace Elements. Tomato plants grown in the glasshouse in any of several soil types amended with sewage sludge have markedly reduced growth and symptoms of chlorosis (Touchton and Boswell 1975). Mineral analysis reveals that Zn content is much higher with sludge application and there are increased levels of several heavy metals.

Herbicides. Tomato sensitivity to 2,4-D has been demonstrated (Robbins and Taylor 1957). Tomatoes exposed to even a mild (five ppm) 2,4-D application have reduced yield, delayed maturity, and changes in fruit quality.

Salt. There is considerable interest in the development of salt tolerance in crop plants. Differences between salt-sensitive and salt-insensitive genotypes of the tomato are being studied comparing Lycopersicon cheesmanii ssp. minor (Hook.) C.H. Mull. from the Galapagos Islands with a commercial cultivar 'VF36' (Rush and Epstein 1976). The former survives in full strength sea water while 'VF36' generally does not withstand more than 50% sea water.

8.16 Comparative Studies with Vegetable Crops

Many research projects involve the simultaneous exposure of several species to the same pollutant treatment allowing direct comparisons among species. Such studies allow much more precise definition of species sensitivity differences but they may not allow interpretation of species differences in terms of environment, growth stage and other interactions. Most research projects could be made more relevant and objective by the inclusion of one or more species and cultivars of known sensitivity to the pollutants under study.

Air Pollutants. Growth of endive, spinach and tomato is suppressed when the plants are grown in ambient air containing oxidants compared with plants grown in ambient air passed through an activated carbon filter (Hull and Went 1952). Responses to F of chard, endive, spinach and romaine lettuce have been compared by exposure of plants continuously for four months (Benedict et al. 1964). All species behave similarly. Top growth is affected only if there is visible leaf injury while root growth is somewhat retarded by any HF treatment. In each case F is translocated to the leaf extremities and does not translocate downward in the plant. The effects of HF exposure on fruiting of several vegetable crops have been studied by controlled exposure in growth chambers (Pack and Sulzbach 1976). The order of decreasing sensitivity is bell pepper, sweet corn, cucumber and pea, with reduced seed production the most common response. Effects on fruiting are considered to be independent of foliar injury.

Trace Elements. Tissue culture techniques have been used to establish toxicity levels for Pb, Hg, Cu and Zn in terms of affecting fresh weight increase of explants of cauliflower, lettuce, potato and carrot (Barker 1972). There is considerable variation in toxic levels among both species and heavy metals. In some cases very low levels stimulate growth and, in lettuce, high Zn sometimes stimulates growth.

Cadmium effects on vegetable crops have been the subject of numerous studies in recent years. Studies of Cd concentration effects on growth of several vegetable crops indicate that beets, beans and turnips are most sensitive with 50% growth reduction at 0.2 μg Cd/ml (Page et al. 1972). The corresponding level for corn and lettuce is 1 μg Cd/ml, for tomatoes 5 μg Cd/ml, and for cabbage 9 μg Cd/ml (Fig. 54, 55). Leaves accumulate Cd and concentrations vary widely among species, increasing with increasing solution Cd level.

Several vegetable crops have also been grown in Cd-treated soils (John 1973). Highest Cd levels in edible plant parts are found in lettuce and spinach leaves, followed by brassica tops, radish and carrot roots and pea seeds. Cadmium toxicity results in growth retardation in some cases. Responses to Hg have been studied similarly (John 1972a). Leaf lettuce, spinach, broccoli, cauliflower, peas, radishes and carrots are grown in $HgCl_2$-treated soil. Root concentrations range from 0.387 μg Hg/g of lettuce roots to 2.447 μg Hg/g of cauliflower roots when 20 μg Hg/g soil is applied. Spinach leaves and radish roots have the highest concentrations of any edible parts.

8.17 Vegetable Composition and Nutritional Quality

Little information is available on the effects of air pollutants on crop composition and nutritional value. Quality assessments of cabbage, carrots, corn, lettuce, strawberries and tomatoes grown in carbon-filtered air with or without added O_3 showed that nutrient quantities are generally unaffected by O_3 exposure (Pippen et al. 1975). Total solids tend to be up in cabbage, down in carrots, corn and tomatoes, and unaffected in lettuce and strawberries (Table 36). Ascorbic acid increases in some plants in response to O_3 exposure. Effects on N, ash and niacin concentrations also vary among species.

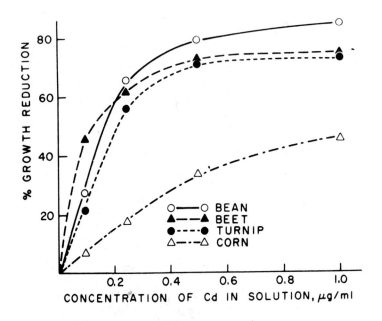

Fig. 54. Growth reduction of bean, beet, turnip and corn in relation to Cd
concentration of nutrient solution culture (Page et al. 1972.
Reproduced from JOURNAL OF ENVIRONMENTAL QUALITY 1, 288-91, by
permission of the American Society of Agronomy, Crop Science
Society of America and Soil Science Society of America).

Table 36. Composition and solids trends in edible parts of vegetable crops
exposed to O_3.

| Species | Solids trend | Number of significant composition effects | | |
		Total	Following solids	Counter to solids
Cabbage	Up	5	5	0
Carrots	Down	1	0	1[a]
Corn	Down	5	2	3[b]
Lettuce	None	2	0	0
Strawberries	None	2	0	0
Tomatoes	Down	5	4	1[c]

[a] Niacin   [b] N, Vitamin C (ascorbic acid), vitamin C (total)   [c] Ash

Data from Pippen et al. (1975) Journal of Food Science Vol. 40, pp. 672-6.
Copyright C by Institute of Food Technologists.

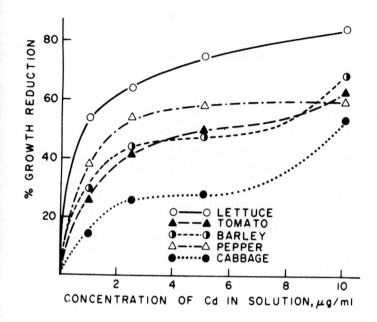

Fig. 55. Growth reduction of lettuce, tomato, barley, pepper and cabbage
in relation to Cd concentration of nutrient solution culture
(Page et al. 1972. Reproduced from JOURNAL OF ENVIRONMENTAL
QUALITY, 1, 288-91, by permission of the American Society of
Agronomy, Crop Science Society of America and Soil Science
Society of America).

CHAPTER 9

FRUIT CROPS

Tree Fruits

9.1 Apple

There have been very few studies of apple responses to air pollutants but many studies of pesticide effects because of the intensive use of pesticides required for successful pest-free production of this species.

Oxidants. Apple fruit responses to high O_3 levels include production of grey to brown pitted areas around lenticels on fruit of three of the five cultivars tested (Miller and Rich 1968). The response is not affected by light or humidity at the levels used. Visible O_3 injury to foliage includes brown stippling and light yellow mosaic patterns (Elfving et al. 1976). Inclusion of young apple trees in chemical protection studies indicated that apple is protected against O_3 injury by the application of foliar diphenylamine spray or dust (Gilbert et al. 1975). An antitranspirant 'Wilt Pruf' applied as a foliar spray also protects (Elfving et al. 1976).

Sulphur Dioxide. Apple trees lose their leaves early in July in regions with high SO_2 concentrations. At lower SO_2 levels, leaves have interveinal necrosis but remain on the trees. Apple cultivars tested in the field with a portable exposure device are sensitive to SO_2 (Spierings 1967b). Apple trees cv Golden Delicious exposed to O_3 and/or SO_2 are relatively sensitive to SO_2 but insensitive to O_3 (Table 37). The mixture of gases results in a greater than additive effect in retarding shoot growth as well as in producing greyish-green water-soaked areas on mid-shoot leaves.

Ammonia. Apple fruit exposed accidentally to NH_3 in storage have a characteristic black spotted appearance of the lenticels (Brennan et al. 1962).

Acid Rain. Experiments have been conducted to determine simulated acid rain injury symptoms on apple (Fig. 56, 57).

Trace Elements. The effects of heavy metals on apple trees are of particular interest because of the use of many metal-containing pesticides in the production system (Table 15). A survey of apple orchard soils reveals that As and Pb levels are elevated as a result of many years' use of Pb and Ca arsenate (Table 38). Duration of pesticide usage and rates of application determine soil concentration. The Hg content in apple orchard soils has also

Table 37. Effects of SO_2 and O_3 alone and in combination on shoot growth, leaf injury, and leaf abscission of 'Golden Delicious' apple trees after 6 hr exposures. Data from Kender and Spierings (1975)

Treatment	pphm	Shoot growth[x] cm	% leaf surface injured[y]	% leaf abscission[z]
Control		23.0	0.0	0.0
O_3	30	23.1	0.0	0.0
SO_2	40	23.5	1.8	2.0
	60	24.3	7.3	5.0
	250	18.7	72.9	62.3
$SO_2 + O_3$	40 + 20	17.5	2.8	5.7
	80 + 40	14.8	5.9	16.8
	150 + 40	14.8	17.2	28.8

[x] 21 days after exposure.

[y] Mean of all leaves 3 to 5 days after exposure.

[z] 10 days after exposure.

Fig. 56. Simulated acid rain injury on apple fruit (Photo courtesy of J.T.A. Proctor).

152

Fig. 57.   Simulated acid rain injury on apple leaves (Photo courtesy of J.T.A.
Proctor).

been elevated by the use of PMA over a 20-year period.   Copper concentrations
in orchards have been raised slightly by application of Cu fungicides over an
80-year period.

Mercury has had considerable study.   This element translocates from apple
foliage to growing fruit and new foliage (Ross and Stewart 1962).   Sprays of
PMA result in little volatilization of Hg but considerable Hg translocation to
fruit which are covered during spraying, as well as translocation to new
terminal growth (Table 39).

This significant translocation of Hg to growing apple fruit has been confirmed by the application of PMA sprays to four apple cultivars (Ross and
Stewart 1962).   The cultivars behave similarly.   Sprays of several other Hg
compounds result in similar translocation of Hg to growing fruit.

Cadmium responses have also been studied (Ross and Stewart 1969).   Follow-
ing cover sprays of $CdCl_2$, Cd accumulates in apple fruit as they mature.
Cadmium residues persist in the foliage throughout the growing season and the
Cd may be translocated from the foliage to the fruit.

Salt.   Salt injury to apple trees is of particular concern for orchards
near roadways on which deicing salt is used.   Increased levels of Na and Cl,
from simulated salt spray, result in suppression of flowering and dieback of

Table 38. Metal, non-metal and pesticide residues in surface soil of 31 apple
orchards located in Ontario, Canada.

| Metals, non-metals and pesticides | Residues in dry soil (ppm) | | | |
	Mean	Range	S.D.	% of soils above normal[y]
As	54.2	3.2-126	25.8	84
Cu	39.0	11.3-110	20.7	26
Hg	0.29	0.03-1.14	0.29	61
Fe	9,250	3,820-21,460	4,435	0
Mn	317	120-861	285	0
Pb	247	6.4-888	207	87
Zn	31.7	5.4-117	310	6
DDE	7.67	0.78-33.3	8.46	100
TDE	1.22	0.04-4.00	1.10	100
opDDT	2.28	0.23-8.90	2.35	100
ppDDT	32.1	1.08-294	63.3	100
DDT	43.3	1.33-334	85.2	100
Dicofol	0.37	ND[z]-2.40	0.60	64
HEOD	0.03	ND-0.38	0.08	45
Endosulfan	0.26	ND-2.63	0.64	45
Endrin	<0.001	ND-0.003	–	9
Heptachlor epoxide	0.001	ND-0.097	0.020	14
Lindane	0.02	ND-0.39	–	5
Methoxychlor	<0.001	ND-0.004	–	5
Ethion	<0.001	ND-0.026	–	5
Parathion	<0.001	ND-0.021	–	5

[y] % of soils above normal background level.

[z] ND - not detected < 0.0004.

Data from Frank et al. (1976b)

Table 39. Mercury content of apple fruit and foliage from limbs protected
by Mylar bags during PMA spraying

Zone	Fruit residue µg Hg/10 apples	Foliage residue µg Hg/100 cm^2
Terminal	26.2	0.4
Intermediate	27.2	0.3
Near bag opening	27.6	0.1
Sprayed area	28.2	5.3
New terminal growth outside bag		5.3

Data from Ross and Stewart (1962)

terminal shoots (Hofstra and Lumis 1975). Symptoms are similar whether on
roadside trees exposed to salt spray or on trees sprayed with 2% salt solutions.
The tissue concentration thresholds for injury are 0.20% Na and 0.50% Cl on a
dry weight basis.

Pesticides. The build-up and persistence of pesticide residues in apple orchards have been documented. The use of Pb arsenates on apple orchards in Ontario, Canada over the past 70 years has raised soil As and Pb from background levels of 7.4 and 6.4 ppm to 54.2 and 247 ppm respectively (Table 38). The use of PMA for 20 years for scab control has elevated Hg levels from 0.08 to 0.29 ppm. Copper used as a fungicide over a 70 year period has increased Cu in the soils from 21 to 39 ppm. Iron, Mn and Zn are not increased by orchard management practices. Among insecticides, the no longer used DDT, TDE and their metabolite DDE have total mean residues in apple orchards of 43.3 ppm. Other organochlorine residues found are endosulfan, endrin, HEOD and methoxychlor. Dicofol, an acaricide, is also present. Among organophosphorus insecticides only ethion and parathion are found, and these only in trace amounts. The distribution of DDT residues in apple trees and herbage in an orchard sprayed annually has been documented (Stringer et al. 1975).

The direct effects of pesticides on apple tree growth and yield are also of concern. Apple trees treated with several pesticide combinations over a three-year period indicate that several pesticides have an unfavourable effect on yield (Donoho 1964). It is recommended that pesticide chemicals be evaluated on fruit and trees prior to general usage.

Several studies have focused on the effects of fungicides on apple tree growth and yield. In comparisons of fungicides, yields are highest with captan followed in order by dodine, PMA-captan and dichlone (Ross and Longley 1963). PMA-captan results in the largest apples. Dichlone causes a rough fruit surface while captan sprays result in the most attractive fruit. There are no differences in apple yields between captan and dodine fungicides (Groves and Rollins 1966). Highest yields occur when zineb is sprayed on young apple trees, followed by captan, dodine, dichlone, ferbam, glyodin, and Bordeaux mixture (Ross et al. 1970). Fruits are large with ferbam treatment although yields are not high.

Pesticide effects on photosynthesis and respiration of apple leaves have been determined (Ayers and Barden 1975). Most chemicals of the 33 tested have no effect. Plictran and formetanate hydrochloride increase photosynthesis. Oil and emulsifiable concentrate formulations generally reduce photosynthesis. The method of treatment has a marked effect on the response to some of the chemicals. Photosynthesis of apple trees is reduced by dicofol, an acaricide, with differences between cultivars (Sharma et al. 1976b). Sprays of dodine do not affect photosynthesis.

Herbicides. The extensive use of herbicides in apple orchards is a cause for concern because their presence in the soil may affect tree growth. Apple responds to soil-incorporated simazine, and a concentration of 2.5 ppm reduces shoot and root growth of 'Cox's Orange Pippin', a sensitive cultivar (Dvorak 1968). Soil-applied herbicides affect the photosynthesis rate of 'Golden Delicious' apple foliage (Sharma et al. 1977a). Monuron and atrazine cause the greatest decrease, followed by simazine, but diuron does not affect photosynthesis. The herbicides decrease photosynthesis by 10 days after soil application and for at least 40 days thereafter. Monuron and atrazine also decrease the plastochron index, an indicator of tree growth. These results are in agreement with the knowledge that atrazine and monuron are relatively mobile and simazine and diuron relatively immobile. The group as a whole, while acting primarily by inhibiting the Hill reaction of photosynthesis, has relatively slow movement in the soil and is absorbed on the cation exchange complex. Such herbicides are therefore considered to be relatively safe for soil application in fruit crop production.

9.2 Apricot

The responses of apricot to F have been characterized with the cultivars 'Chinese' and 'Royal' rated as sensitive to F and 'Moorpark' and 'Tilton' as intermediate (Jacobson and Hill 1970). Relative sensitivities to other pollutants have not been established.

Fluoride. The characteristics of F injury to apricot in the vicinity of an Al reduction plant include leaf injury, defoliation, and bud injury, resulting in severe yield reduction in subsequent years (De Ong 1946). Exposure of 'Tilton' apricot to HF gas decreases pollen growth in stylar tissue, while pollen germination is unaffected in NaF-containing synthetic media (Facteau and Rowe 1977). In contrast, exposure of similar plants to HCl gas has no effect on pollen tube growth over a wide range of concentrations and times.

9.3 Avocado

Avocado seedlings exposed for a period of eight weeks to the reaction products of O_3 and 1-hexene have sharp reduction in stem elongation, leaf expansion and plant weights (Taylor et al. 1958). Injury symptoms include a brown discolouration on the lower surface of young expanding leaves, together with marginal and tip necrosis on some leaves.

9.4 Cherry

Among the common air pollutants, only an intermediate sensitivity to F is recorded for cherry (Jacobson and Hill 1970). Decreased cherry pollen germination and pollen tube growth are related to increasing F levels (Facteau et al. 1973). This observation may explain reduced fruit set in cherry exposed to HF. Dose rate and F residue in the flowers are directly correlated.

9.5 Citrus

Citrus species responses to air pollutants have been studied relatively extensively, especially in California, U.S.A., where production areas are subject to phytotoxic air contaminant episodes.

Oxidants. Exposure to the reaction products of O_3 and 1-hexene significantly reduces the growth of rooted 'Lisbon' lemon cuttings as well as 'Lisbon' lemon scions growing on citrange rootstocks (Taylor 1958). Grapefruit seedlings respond similarly. No visible symptoms of leaf injury are observed, but the rate of water use is reduced by as much as 25%. If exposure is extended, there is premature senescence and drop of older leaves.

These observations have been extended by the use of filtered and non-filtered plastic greenhouses to study ambient air pollutant effects on water use and apparent photosynthesis of citrus trees (Thompson et al. 1967). The ambient mixture of pollutants in California, U.S.A., reduces the rate of water use and apparent photosynthesis, compared to filtered air. Attempts to determine the separate effects of components of the ambient smog complex have not been successful. The same facilities can be used to show that the photochemical smog complex reduces the yield of both lemon and navel orange trees as well as reducing the rate of water use and apparent photosynthesis (Thompson and Taylor 1969). Leaf drop, and in navel orange, fruit drop, are less in filtered air than in ambient air. The fruit yield is reduced by 50% in ambient air in some cases, yet the trees show no easily recognizable injury symptoms.

Reduced fruiting of citrus due to oxidants is considered to be a cumulative response in which reduced water uptake and photosynthesis combine with leaf drop to decrease tree vigour (Thompson 1969; Thompson and Taylor 1969).

Studies of O_3 effects, in contrast to total oxidant effects, have also been reported for navel oranges (Thompson et al. 1972). Fruit drop increases with increased O_3 and number and weight of mature fruit decrease (Table 40). The entire photochemical smog complex reduces yields much more than does O_3 alone.

Exposure of young navel orange trees or branches to PAN results in growth retardation with a trend toward reduced yield of mature fruit, but short-term apparent photosynthesis is not affected by rates of 20 - 80 ppb of PAN (Thompson and Kats 1975).

Navel oranges have premature fruit drop at an early stage of development as one of the effects of oxidant smog. Dusting on the two antioxidants DPPD and NBC decreases fruit drop and significantly increases fruit yields (Thompson and Taylor 1967).

Table 40. Leaf and fruit drop of navel orange trees in response to O_3.

	Leaf drop kg/tree	Fruit set na/tree	Fruit drop na/tree	Fruit drop %	No. fruit/ tree	Fruit wt kg/tree
C-filtered air	5.11	897	452	50.8	445	81.1
C-filtered air + ½ ambient O_3	5.13	848	441	52.9	406	77.0
C-filtered air + ambient O_3	7.55**	792	487	62.7*	304*	52.6*
Ambient air	5.69	547**	505	76.1**	142**	28.5**
Outside checks	5.28	435**	334	79.6**	101**	24.0**

*,** Significantly different from C-filtered air treatment at 5%, 1% levels.
Data from Thompson et al. (1972). Reprinted with permission from Environ. Sci. Technol., 6, 1014-6. Copyright by the American Chemical Society.

Evaluation of NO_2 responses of five citrus cultivars indicates that exposure to high concentration-short time dosages results in rapid tissue collapse, necrosis, and defoliation (MacLean et al. 1968). There are cultivar differences and older tissue is more resistant than young. Navel orange trees are severely defoliated by 35 days of 50 or 100 pphm NO_2, while exposure to 25 pphm and lower levels causes increased leaf drop and reduced fruit yield (Table 41).

Table 41. Nitrogen dioxide exposure and ambient air pollutant effects on leaf drop and yield of navel oranges.

	Leaf drop kg/tree	Fruit drop %[z]	Yield of fruit/tree	
			No.	kg
Filtered air	6.55	71.9	346	75.7
Ambient air	7.77	91.4	85	20.4
Outside check	6.69	90.6	81	19.6
Filtered air + 6.25 pphm NO_2	7.24	88.1	135	32.7
Filtered air + 12 pphm NO_2	7.87	82.0	218	51.8
Filtered air + 25 pphm NO_2	8.59	85.2	156	40.0
Filtered air + 50 pphm NO_2	-	93.7	67	15.3
Filtered air + 100 pphm NO_2	-	90.2	94	22.9

[z] % of total number of fruit set.
Data from Thompson et al. (1970)

Sulphur Dioxide. Orange trees are insensitive to concentrations of SO_2 expected in the field in California, U.S.A. (Darley et al. 1956).

Fluoride. Fluoride accumulation in citrus foliage in California, U.S.A., has been surveyed (Kaudy et al. 1955). Leaves from industrialized areas have up to 211 ppm F even though there are no visible injury symptoms. Leaves in many non-industrialized areas have one ppm F or less. A survey conducted in Florida, U.S.A. indicates that unique chlorosis patterns occur on citrus leaves exposed to F emitted from superphosphate manufacturing plants (Wander and McBride 1956).

Controlled exposures to HF gas indicate that prolonged exposure is deleterious to a wide range of citrus species but there are species differences in severity of injury (Brewer et al. 1960a). Decreased growth in the presence of HF is greater than can be accounted for by leaf necrosis. Decreased leaf size and increased foliar necrosis are the two most common symptoms of increasing exposure to HF. More detailed studies on the navel orange have indicated that this species is sensitive to atmospheric HF concentrations as low as two to three ppb (Brewer et al. 1960b). Growth and vigour of young trees are reduced without visible injury symptoms. There is a 25 to 35% reduction in average leaf size with HF exposure, but this may be due to heavy leaf drop in early summer which stimulates growth of many new smaller leaves.

Five citrus cultivars exposed to high concentration – short term dosages of HF have tip, marginal and intercostal necrosis, and leaf abscission (MacLean et al. 1968). Cultivars differ in sensitivity and young tissue is more sensitive than old.

The similarity of effect of HF gas and NaF sprays has been shown by spraying 'Washington' navel oranges with NaF periodically over a six-year period (Brewer et al. 1967a). Plants treated with NaF have toxicity symptoms similar to those on plants exposed to HF gas, including interveinal chlorosis, premature leaf drop, reduced leaf size, and reduced fruit production. No effects on fruit quality are noted and F translocation to fruit is minimal when early season applications of NaF are made. Further studies of NaF solution spray effects have shown their similarity to HF exposures (Brewer et al. 1969c). Weekly sprayings of young fruit trees with NaF or HF solutions result in very similar symptoms to those from intermittent exposure to HF gas. Continuous exposure to HF gas results in more severe growth retardation than does intermittent exposure to higher concentrations or periodic spraying with soluble fluorides.

'Valencia' orange trees subjected to twice daily drenching showers or daily fine water sprays accumulate less F than unsprayed trees (Brewer et al. 1969a). The drenching showers are more effective on older plants than spraying, although their effect is similar on younger plants. Both treatments result in reduced growth suppression by F in proportion to decreased F uptake.

An effective chemical protection treatment against F injury is available (Brewer et al. 1969b). Applications of Ca(OH)$_2$ sprays to commercial 'Valencia' orange trees improve fruit production in the presence of severe F pollution. Sprays of CaCl$_2$ are not beneficial even though the accumulation of F in citrus tissue is about the same for the two salts.

Studies of citrus seedling exposure to SO$_2$, HF or mixtures of the two pollutants indicate that mixture effects are additive in reducing growth and leaf area of 'Koethen' sweet orange but there is no leaf necrosis (Matsushima and Brewer 1972). 'Satsuma' mandarin orange has leaf chlorosis and abscission in the mixed gases but growth is less affected.

9.6 Peach

Peach is recognized as sensitive to F and ethylene, intermediate in sensitivity to 2,4-D, Cl$_2$ and Hg vapour, and insensitive to H$_2$S (Jacobson and Hill 1970). Fluorides induce an unusual disorder in peach fruit known as "suture red spot" or "soft suture" (Benson 1959). Tissue on one or both sides of the suture ripens prematurely and splitting occurs along the suture. The affected areas are soft and may be decomposed by harvest time. The severity of this disorder can be reduced by application of lime sprays or dust to foliage. Peach fruit held in storage may be accidentally injured by NH$_3$ fumes released from the refrigeration system (Brennan et al. 1962). The symptoms on some cultivars are discrete spots while on others they are large brown bruise-like patches.

Peach seedling sensitivity to Al in nutrient solution has had detailed study including diffusive resistance measurements and stomatal and root observations (Horton and Edwards 1976). Diffusive resistance increases with increasing Al but this is associated more with decreased root volume than with stomatal characteristics.

Small Fruits

9.7 Blueberry

Very limited observations on this species include its rating as F-sensitive (Jacobson and Hill 1970). Highbush blueberry responses to SO$_2$ have been evaluated (Brennan et al. 1970). The four cultivars tested vary in response but are generally quite insensitive compared to sensitive species like begonia and zinnia. Young foliage is more sensitive than old.

9.8 Currants

The SO$_2$ sensitivity of currants is in dispute. The species is rated as SO$_2$ insensitive by Jacobson and Hill (1970) but Guderian (1969) has found currants to be very susceptible to SO_2 injury. Injury symptoms, including abscission of many leaves at an early stage, are evident at concentrations of 40 pphm and higher.

9.9 Grape

Grape is categorized as sensitive to O_3 and 2,4-D and as intermediate in sensitivity to SO_2 and Cl_2 (Jacobson and Hill 1970).

Oxidants. Chlorotic stipple on grape leaves as a response to atmospheric oxidants was first reported in California, U.S.A. (Richards et al. 1958). The upper surface dark brown to black punctate spotting of grape leaves is attributed to the presence of O_3 in vineyards. Growing grapes in ventilated plastic greenhouses in the same area and using activated carbon filtration of incoming air to reduce O_3 levels to about 25% of ambient results in an increase in fruit size and quality, and the stippled chlorosis of leaves is markedly reduced (Thompson et al. 1969). Tissue injury is associated with loss of chlorophyll which, in turn, reduces photosynthesis, resulting in smaller berries, lower juice content, and reduced vine growth.

A grape leaf disorder called "brown leaf" with symptoms resembling those noted in California, is found in the Great Lakes region of the U.S.A. (Shaulis et al. 1972). The disorder had been seen for many years without identification of the real cause. The disorder is characterized by small dark brown interveinal flecks on the upper surfaces of mature leaves, and premature senescence and abscission of leaves (Fig. 58). Experimental exposures are used to prove

Fig. 58. Brown leaf disorder of grape leaves associated with ambient O_3 episodes (Author's photo).

that O_3 causes the stipple. Identical symptoms are obtained by exposing two-
year old potted grape vines to 30 or 60 pphm O_3 for six hr. The relative
susceptibility of a few cultivars has been determined by controlled exposures
and a large number of grape cultivars growing outdoors in New York State,
U.S.A., and Ontario, Canada, have been rated for oxidant injury and a wide
range of sensitivity found (Kender and Carpenter 1974). Oxidant stipple
ranges from very mild to severe on 53 American grape cultivars and very mild
to moderate on 41 French hybrid grape cultivars. No consistent genetic
relationships can be detected and symptom severity varies according to
localized growing conditions.

Chemical protection experiments have been conducted by applying benomyl
to grapes (Kender et al. 1973). Reduced oxidant injury is found if three or
more applications are made during the growing season. The degree of protection
is directly proportional to the frequency of benomyl application. The relationship of management practices to oxidant injury has also been described (Kender
and Shaulis 1976). 'Concord' grapevines have less oxidant injury with added
N compared to no added N, with own-rooted vines compared with those grafted
on Couderc 3309 rootstock, and with flower cluster thinning compared to no
thinning (Table 42). Pruning severity and training systems do not affect
oxidant injury.

Sulphur Dioxide. The SO_2 response of grapes has been studied (Fujiwara
1970). There is a decrease in shoot growth and cluster formation along with
leaf necrosis and leaf drop after long exposure to 13 or 26 pphm SO_2 (Table 43).

Fluoride. The pattern of F accumulation by grapes growing downwind from
a steel mill has been established (Brewer et al. 1957). There is a gradual
accumulation of F in the leaves during the growing season. Leaf injury,
consisting of marginal necrosis, generally is well correlated with F content
but the extent of injury is apparently influenced by temperature, humidity
and soil water. The leaves have a much higher F content than the fruit.

Trace Elements. Accumulation of Pb in grapes grown in the vicinity of
roadways has been studied (Favretto et al. 1975). Lead content decreases
rapidly from the roadway to about 100 m distance.

9.10 Strawberry

There has been little research on pollutant response of this species
which is surprising in view of its widespread production in surburban areas
where pollution problems would be expected. It is rated as insensitive to
F and H_2S (Jacobson and Hill 1970) and insensitive to O_3 and SO_2, separately
or together (Rajput et al. 1977). The high insensitivity to O_3 and/or SO_2

Table 42. Oxidant stipple of 'Concord' grapevines as influenced by N fertilization (N), rootstock (R), pruning severity (P), weed control
(W), and flower cluster thinning (T).

		% oxidant stipple[y]			
		1971	1972	1973	3-yr ave.
N	N 0 kg/ha/year	20**	73**	48**	47**
	N 56	15	59	35	36
	N 112	14	53	32	33
R	Own	7	48	22	26
	Couderc 3309	27**	75**	55**	52**
W	Sod	20	64	40	41
	Cultivated	14	61	37	37
P	Recommended scale	18	63	40	40
	Light scale	16	60	37	38
T	None	19**	66**	–	43**[z]
	Thinned	15	59	–	37
Interactions					
	WXT	**	*	*[W]	*
	RXT	**	*	*[W]	*
	NXT	*	–	–	*

*, ** Statistically significant at 5% and 1% levels.

[y] Percentage of the leaf area at nodes 1-6 that exhibited oxidant stipple.

[z] Residual effects of flower cluster thinning.

Data from Kender and Shaulis (1976)

Table 43. Severity of leaf lesions and leaf fall of 'Fredonia' grape under
SO_2 exposure.

No. of days from start of exposure	Severity of leaf necrosis (leaf fall rate %)			
	Control	6.5 pphm	13 pphm	26 pphm
22	–[z]	–	3.2 (0)	21.7 (7.8)
28	–	–	6.2 (2.8)	27.5 (15.0)
37	–	–	16.5 (7.2)	66.4 (41.0)
56	–	15.2 (9.8)	73.8 (51.0)	81.6 (72.7)
70	–	24.5 (12.0)	70.3 (51.4)	86.6 (74.3)

[z] No visible injury.

Data from Fujiwara (1970)

occurs in many cultivars. The only mild leaf injury symptoms obtained are in some cultivars exposed to substantial levels of the combined gases. No growth and development effects of O_3 and/or SO_2 can be noted.

Strawberry plants exposed to HF have fruit deformities in proportion to the concentration of gas with slight apical deformation at rates as low as 0.55 µg F/m^3 (Pack 1972). Deformations result from inhibition of seed development and are independent of visible injury to the foliage (Table 44).

Table 44. Flowering, fruiting, and F accumulation by strawberry plants as influenced by continuous long-term (5 mo) exposure to HF.

	Marshall						Columbia	
	0.55		2.0		10.4		10.4 µg F/m^3	
	Control	HF	Control	HF	Control	HF	Control	HF
Flower no.	866	955	908	967	578	655	634	577
Flowers to fruit %	64.4	60.8	61.5	56.3	69.0	53.2*	68.6	48.1*
Fruit wt. g	6.61	6.73	6.69	6.14*	7.53	3.92	7.25	2.58**
Development rating	5.3	4.5*	5.4	4.0**	5.3	2.5**	5.0	2.2**
F content ppm								
Leaves	1.6	110	3.3	340	4.0	2500	2.5	2200
Fruit	0	2.5	2.0	4.5	3.0	100	3.0	72

*,** Significantly different from control at 5%, 1% levels.

Data from Pack (1972)

9.11 Tea

Very little information is available on pollution responses of tropical and sub-tropical horticultural species. Such information will be needed more in the future as many less-developed countries are undergoing urban and industrial development. For example, commercial tea plants may accumulate F and wide ranges of F concentration in tea have been reported (Singer et al. 1967).

CHAPTER 10

FLOWER CROPS

There have been many reports of flower crop sensitivity to air pollutants but few on responses to other forms of pollution. Air pollutant effect reports have resulted from observations and experiments in controlled environment facilities, under outdoor conditions, or in glasshouses. The evaluation of cultivar sensitivity has revealed that a wide range exists in most species.

10.1 Begonia

This species is more sensitive to O_3 or SO_2 than many other flowering bedding plants (Adedipe et al. 1972a). Some cultivars are sensitive to the concentrations of these gases found outdoors (Leone and Brennan 1969). Red cultivars are more tolerant than white, and the sensitivity to O_3 or SO_2 increases with increasing atmospheric moisture (Table 45).

Table 45. Ozone, SO_2 and relative humidity effects on foliage injury of Begonia semperflorens.

	Relative humidity (%)						
	35 - 49				65 - 74		
	O3 concentration (pphm)						
	10	15	20	25	10	15	20	25
Cinderella white	0[z]	0	0	Sl	0	Sl	S	S
Scandinavian pink	0	0	0	M	0	Sl	M	S
Scandinavian red	0	0	T	Sl	0	M	M	S
Cinderella rose	0	0	Sl	Sl	T	Sl	S	S
Scandinavian white	0	0	Sl	M	T	M	S	S
Red comet	T	T	Sl	Sl	Sl	M	M	S
White comet	T	Sl	S	S	Sl	S	S	S

| | SO_2 concentration (pphm) | | | | | | |
	80	135	160	215	42	51	70	80
Scandinavian pink	0	0	0	0	0	0	0	Sl
Scandinavian white	0	0	0	0	0	0	0	Sl
Scandinavian red	0	0	0	0	0	0	0	M
Cinderella rose	0	0	T	Sl	0	0	Sl	M
Cinderella white	0	0	T	Sl	0	0	Sl	Sl
White comet	0	0	Sl	Sl	0	M	S	S

[z] Degree of injury: 0 = none, T = trace, Sl = slight, M = moderate, S = severe.

Data from Leone and Brennan (1969)

The separate and combined effects of O_3 and SO_2 on Rieger begonia have been studied (Gardner and Ormrod 1976). There is much more leaf injury when the two gases are combined than when plants are exposed to separate gases. Threshold concentrations for leaf injury are about 20 pphm O_3, 120 pphm SO_2, or a combination of greater than about 10 pphm O_3 and 60 pphm SO_2 during midday for five days (Table 46).

Table 46. Ozone and/or SO_2 effects on % of Rieger begonia leaves injured 2 weeks after 4 hr exposure per day for 5 days.

Treatment		Pollutant concentration (pphm)								
	O_3	10	10	10	20	20	20	30	30	30
	SO_2	60	120	180	60	120	180	60	120	180
Control		$0a^z$	0a	0a	0a	2a	0a	1a	2a	2a
O_3		0a	1a	2a	32b	23b	21b	37b	42b	42b
SO_2		0a	22b	67b	4a	21b	33b	3a	48b	48b
$O_3 + SO_2$		39b	44c	67b	67c	50c	50c	32b	62c	62c

^{z} Mean separation within colums by Tukey's test, 5% level.
Data from Gardner and Ormrod (1976)

10.2 Carnation

While carnation is rated as sensitive to O_3 (Jacobson and Hill 1970), the main concern of producers is its high ethylene sensitivity. Ethylene-induced "sleepiness" or premature senescence of flowers can be a serious problem in commercial carnation production and cut flower distribution.

Oxidants. The occurrence of a rapidly-developing tip burn disorder of carnations is related to ambient oxidant concentrations (Feder et al. 1967). Ozone or a mixture of pollutants is implicated as a causal agent. The amount of flowering of carnations is adversely affected by exposure to about 7.5 pphm O_3 for about 10 days at the time of bud initiation (Feder and Campbell 1968). Eight weeks of this treatment are required for leaf tip burn symptoms to appear. Carnation plants exposed to 10 pphm O_3 for one to three months have increased internode length along with smaller leaves with progressive destruction of leaf tissue (Feder 1970).

Ethylene. The adverse effects of illuminating gas and ethylene on carnations have been thoroughly studied (Crocker and Knight 1908). This species is extremely sensitive in terms of bud kill and prevention of bud opening. The

toxicity of illuminating gas is a function of the ethylene it contains.
Carnation responses to ethylene have also been characterized by exposing
flowers to ethylene then returning them to ethylene-free air (Nichols 1968).
There are no visible responses to 20 pphm for six hr but irreversible wilting
at 20 pphm for 48 hr. Carbon dioxide accumulation during ethylene treatment
delays senescence. Carnation flower life thus can be prolonged by high CO_2
in the presence of low concentrations of ethylene (Smith and Parker 1966;
Uota 1969). Insensitivity to ethylene is doubled by 10% and quadrupled by
20% CO_2.

Effects of duration of ethylene exposure and temperature have been determined by developing a dosage term, ppb-hr, to relate concentration and exposure
time (Barden and Hanan 1972). For example, open flowers have no keeping life
at $1.7^{o}C$ with 47,000 ppb-hr ethylene (Table 47). The corresponding dose at

Table 47. Ethylene dosage levels (ppb-hr) required to obtain 100% and 20% loss
of carnation cut flower keeping life.

	Temperature ^{o}C	Dosage required for loss in keeping life[z]	
		100%	20%
Open flowers	21.1	4,400	2,100
	10.0	10,000	4,400
	1.7	47,000	14,500
Buds	21.1	8,400	3,100
	10.0	25,000	14,000
	1.7	55,000+	55,000+

[z] Keeping life based upon a control group of flowers subjected to C_2H_4 levels
below 10 ppb.

Data from Hanan (1973b)

$21.1^{o}C$ is 4,400 ppb-hr. Buds are less susceptible than flowers. Holding
blooms at cool temperature ($5^{o}C$) prior to exposure reduces the injury threshold
to 1/4 to 1/2 that of flowers held continuously at $20^{o}C$ (Uota 1969). Temperature effect studies on carnation response to ethylene reveal that the sensitivity increases sharply with increasing temperature, with the threshold concentration dependent on prior stresses to the flowers (Maxie et al. 1973).
Those subject to high temperature, long-term storage, or low concentrations
of ethylene have increased sensitivity to ethylene.

Observations at Denver, U.S.A. indicate that most ethylene-induced injury
to carnation flowers in urban areas may occur after the flowers are purchased
by the consumer and displayed at room temperature, rather than in production or

refrigerated storage (Hanan 1973b). Carnations in a flower shop located at a traffic light-controlled road crossing never opened and were thus unsaleable, presumably as a result of ethylene injury (Ten Houten 1974).

Carnation plants exposed to long-term, low concentrations of ethylene during rapid elongation are more susceptible to stem growth inhibition than are plants treated at stages of less rapid growth (Piersol and Hanan 1975). Stems on treated plants are not only shorter but have more lateral branches and shorter leaves even at the low ethylene concentrations occasionally found in urban areas (Fig. 59). Some carnation flowers exposed to ethylene develop early wilting symptoms which later disappear unless the plants are exposed to even higher doses of ethylene in which case the wilting is irreversible (Mayak and Kofranek 1976). Carnation flower responses to ethylene exposure have been further characterized (Mayak et al. 1977). Water uptake decreases rapidly about four hr after exposure, followed by inrolling of petals about two hr later. Ionic leakage is also enhanced, coinciding with the decreased water uptake. Carbon dioxide inhibits the petal inrolling and decline in water uptake.

Fig. 59. Effect of continuous ethylene treatment from time of planting to anthesis on carnation growth (Piersol and Hanan 1975).

Protection. A combination of cytokinins, either kinetin or isopentyl adenine, with 5% sucrose reduces sensitivity and increases flower longevity of carnations exposed to ethylene (Table 48).

Table 48. Cytokinin and/or sucrose pretreatment effects on cut carnation sensitivity to ethylene.

| | Longevity (days)[y] | | | |
| | Ethylene dosage (ppm-hr) | | | |
	0	21.4	42.5	Means
Water	2.9	1.9	1.0	1.9
Sucrose (5% w/v)	5.3	1.1	1.0	2.5
Kinetin (0.23 μM)	3.5	1.1	2.3	2.3
Isopentyl adenine (0.23 μM)	1.6	1.0	1.0	1.2
Kinetin + sucrose	7.6	4.5	5.1	5.8
Isopentyl adenine + sucrose	6.6	3.8	3.1	4.5
Means	4.6	2.2	2.3	

[y] Number of days after termination of the ethylene treatment.

[z] LSD 1% for individual mean separation = 1.52.

Data from Mayak and Kofranek (1976)

Another possible protectant is $AgNO_3$ as a flower spray or dip (Halevy and Kofranek 1977). It protects flowers from ethylene injury but the $AgNO_3$ causes petal spotting.

Management practices may affect ethylene sensitivity (Mayak and Kofranek 1976). Sensitivity in terms of residual flower life is increased by more frequent irrigation and by a longer cool storage period. Storage itself for up to ten days does not affect longevity but stored flowers are more sensitive to ethylene.

10.3 Chrysanthemum

Chrysanthemum is sensitive to O_3 but insensitive to PAN (Jacobson and Hill 1970). Cultivars exhibit a wide range of O_3 sensitivity (Table 49). Ozone injury is characterized by chlorotic leaf stipple (Fig. 60).

Symptoms characteristic of PAN injury consisting of bronzed or glazed leaves have been noted on glasshouse-grown chrysanthemums in New Jersey, U.S.A. (Brennan and Leone 1969). There are cultivar differences in susceptibility and more rapid vegetative growth is conducive to greater injury. Chrysanthemums are quite insensitive to controlled exposure to O_3 or SO_2, so it is considered that the plant injury under ambient conditions is caused by either PAN or some photochemical component other than O_3 (Brennan and Leone 1972). However,

Table 49. Chrysanthemum cultivar sensitivity to 60 pphm O_3 for 3 hr.

Insensitive (0-1.5)[y]

Ann Ladygo	Spinwheel[z]	Golden Yellow Princess Ann
Bright Yellow Tuneful	Yellow Moon	Jessamine Williams
Dark Yellow Tokyo[z]	Larry	Yellow Jess Williams
Dolli-ette	Mermaid[z]	Red Dessert
Golden Cushion	Queens Lace	Rosey Nook
Fuji Jess Williams	Redskin	

Slightly sensitive (1.6-3.0)

Bonnie Jean	Golden Peking	White Grandchild
Distinctive	Lipstick	Wildfire[z]
Flair	Mandalay[z]	
Fuji-Mefo	Ruby Mound	

Sensitive (3.1-4.5)

Baby Tears	Gay Blade	Sleighride
Chris Columbus	Pancho	Tinkerbell[z]
Corsage Cushion	Penguin	Touchdown
Crystal Pat	Pink Chief	Yellow Supreme

Very sensitive (4.6-7.0)

King's Ranson	Minn White	Red Mischied
Mango	Mt. White	Tranquility

[y] Ave leaf injury index (scale 1-10) observed on the upper surface of the
oldest leaves.

[z] Denotes glasshouse cultivars.

Data from Klingaman and Link (1975)

exposure of chrysanthemums to PAN indicated high insensitivity suggesting that
PAN-like symptoms observed outdoors may be due to some other pollutant or the
interaction of PAN with some other pollutant (Wood and Drummond 1974).

In chemical protection experiments, biweekly applications of benomyl or
thiophanate ethyl, or a single application of ancymidol, to an O_3-sensitive
chrysanthemum cultivar provide O_3 protection throughout the growing period
(Table 50). Ancymidol delays anthesis and reduces flower count. Triarimol,
Folicote and SADH are ineffective as O_3 protectants.

Symptoms of ethylene injury to chrysanthemums include epinasty, shortened
internodes, thickened stems, and loss of apical dominance (Tjia et al. 1969).
Many short axillary shoots develop, each with a few small leaves (Fig. 61).

Fig. 60. Ozone injury on upper surface of chrysanthemum leaf (Author's photo).

Table 50. The influence of chemical application and ambient oxidant levels on leaf injury, flower number, and plant fresh wt of 'King's Ransom' chrysanthemum in the glasshouse.

Treatment (ppm)	Leaf injury index[y]	Flower number	Fresh wt (gm)
Benomyl			
0	5.8 bc[z]	26.3 a	287.1 ab
250	1.3 f	26.1 a	247.9 d
750	0.5 g	26.9 a	282.3 abc
Thiophanate ethyl			
0	5.1 cd	26.5 a	228.7 d
250	2.7 e	25.9 a	247.4 d
750	0.9 fg	27.4 a	251.3 cd
Triarimol			
0	5.8 bc	27.3 a	247.9 d
30	6.0 b	26.6 a	240.4 d
90	7.0 a	24.1 ab	256.4 bcd
Ancymidol			
0	4.8 d	23.8 ab	254.6 bcd
4	1.3 f	27.1 a	186.3 e
12	0.4 g	19.9 b	148.2 e

[y] Values are the mean leaf injury index of 8 replications. Zero indicates no visible injury on the upper surface of the oldest leaves while 10 indicates injury on 100% of these leaves.

[z] Mean separation in columns by Duncan's multiple range test at the 5% level.

Data from Klingaman and Link (1975)

Fig. 61. Apparent ethylene injury on chrysanthemum shoots (upper) compared
with normal shoots (Author's photo).

Flowering is impaired; continuous exposure results in no flowering while intermittent exposures result in only crown buds occasionally developing.

Excess trace elements for a horticultural crop like chrysanthemum are of
interest because reclaimed sewage water could be used in glasshouse production
as a source of nutrients in hydroponic crop culture. Investigations of trace
element toxicity to chrysanthemum using 10^{-6}, 10^{-5} and 10^{-4}M solutions of Cu,
Co, Cd, Zn, Ni or Cr show that growth reductions of about 70% result from
feeding 10^{-4}M Cd, Cu or Cr, with less reduction by Co, Ni and Zn (Fig. 62).
Tissue concentrations increase with increasing application rates, with most
minerals generally in the roots followed by the leaves and stems. Cadmium
decreases Mn and increases Zn concentrations in some plant parts.

Chrysanthemums receiving limited applications of domestic liquid sewage
sludge have elemental composition similar to that in plants receiving inorganic
fertilizers (Kirkham 1977b). With increasing application Fe, Cu and Zn

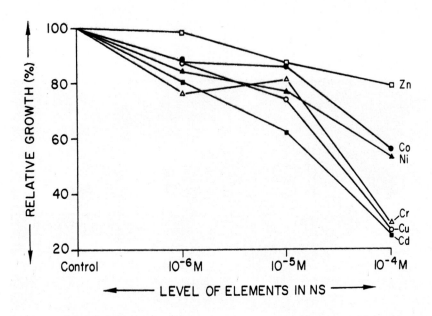

Fig. 62.   Total relative growth in relation to levels of Zn, Co, Ni, Cr, Cu
and Cd in nutrient solution (Patel et al. 1976).

increase in leaves, with the amount depending on the rooting medium.   The
sludge-treated medium has a higher pH and extractable Cu content but lower
extractable K.

10.4 Dracaena

Foliar necrosis of Dracaena cuttings has been investigated (Poole and
Conover 1974). The tissue F content is highly correlated with the severity
of the disorder. The propagating medium influences F uptake and degree of
foliar chlorosis, with the use of unamended peat resulting in the highest
tissue F content.

10.5 Freesia

Freesia is very sensitive to F (Spierings 1969a) and the species has been
suggested as an indicator plant. Freesia has a marginal necrosis in response
to F in contrast to the usual ivory-coloured leaf tip necrosis found on other

monocotyledonous ornamental plants. Freesia plants growing at low HF concentrations can be used to demonstrate the synergistic action between HF and the causal agent of a leaf necrosis symptom (Wolting 1975). The leaf necrosis develops earlier and is more severe in the presence of HF.

10.6 Geranium

The air pollutant responses of this species have had relatively little attention. Geranium plants exposed daily to 10 pphm O_3 for one to three months have reduced leaf size, increased internode length, progressive destruction of leaf tissue and eventual defoliation (Feder 1970). Necrotic O_3-injured tissues of geranium leaves are easily infected by Botrytis cinerea (Manning et al. 1970). Geranium leaves exposed to SO_2 develop necrosis in relation to the initial stomatal opening which is increased by higher atmospheric humidity or removal of CO_2 from the atmosphere (Bonte and Longuet 1975). Exposure to SO_2 is followed by a temporary stomatal closure. Ethylene injury in geranium is characterized by stunting of growth (Fig. 63).

Fig. 63. Apparent ethylene injury on geranium including short internodes and cupped leaves (Author's photo).

10.7 Gladiolus

This species is highly insensitive to O_3 (Ormrod, unpublished data) but very sensitive to F. The high sensitivity of some gladiolus cultivars to F makes this species an important indicator plant of F pollution. Gladiolus responses to F have been detailed in several reports. Fluoride-induced leaf necrosis of gladiolus is confined to outdoor areas with F pollution problems and can be duplicated by controlled exposures to HF (Johnson et al. 1950). Cultivar sensitivity ranges widely, and of 72 cultivars tested, seven were considered insensitive and 13 very susceptible. Insensitive cultivars generally contain more F in the leaf tissue than do the susceptible ones. Thorough descriptions of F injury to gladiolus have been presented (Hitchcock et al. 1962). There are differences in tip burn and F accumulation among cultivars, cultural sites, plant ages and leaf ages. Degree of visible injury can be related to growth and development parameters (Brewer et al. 1966). Flower size and weight, and number of florets per spike are reduced as the severity of F injury increases, as are corm size and weight, and total top weight.

The use of 'Snow Princess' gladiolus as a sensitive indicator of F pollution has been quantified by monitoring both air F content and extent of injury (Spierings 1967a). The average concentration of HF in the atmosphere, the length of leaf tip burn, and the F content of leaf tips are positively correlated. Such determinations provide essential background for the interpretation of F responses in gladiolus and for the use of gladiolus as an indicator species for F pollution.

Determination of sensitivity to F of 110 gladiolus cultivars indicates that there is a wide range of sensitivity with larger leaved cultivars relatively insensitive, and those with purple, lavender or red flower colour less sensitive than those with yellow, white or orange colour (Hendrix and Hall 1958; Ross et al. 1968).

Investigations of the relationship between leaf injury and respiration indicate that there is a great stimulation of respiration in the chlorotic tissue next to necrotic tissue after 2.4 ppb F exposure for 25 days (McNulty and Newman 1957). Respiration is stimulated up to five cm away from the dead area.

Cut gladiolus flower stalks may be injured by fluoridated drinking water (Spierings 1969b). Placing cut leaves or flowers with their bases in F-containing water results in injury to the leaf tips and bracts.

Humidity during exposure to HF affects the severity of F-induced injury (MacLean 1973). Necrosis is more severe on plants exposed at 80% relative humidity than at 65 or 50% (Table 51).

There are mineral nutrition effects on F sensitivity of gladiolus (McCune et al. 1967). Levels of the major nutrients do not affect F level in the foliage but K, P and Mg deficiencies increase tip burn and Ca and N deficiencies

Table 51. Severity of HF-induced necrosis on 'Oscar' gladiolus leaf blades as
affected by relative humidity and leaf age.

Relative leaf age	Relative humidity (%)		
	50	65	80
5 (youngest)	$32.5^y a^z$	29.4a	$41.7b^x$
4	40.9a	39.9a	$48.2a^x$
3	29.2b	33.9a	37.2c
2	6.4c	8.3b	12.8d
1 (oldest)	3.9c	6.2b	$14.4d^x$

[x] Significantly greater than corresponding values at other humidities.
[y] % of leaf length necrotic.
[z] Mean separation within columns at the 5% level.
Data from MacLean (1973)

decrease tip burn in some cultivars. Gladiolus are protected from F-induced
leaf scorch by the application of lime dusts or sprays to the foliage
(Allmendinger et al. 1950), and by $Ca(OH)_2$ sprays (Brewer et al. 1969b).

10.8 Lily

Easter lilies exposed to ethylene exhibit twisting and curling of leaves,
shortening of peduncles, premature whitening of flower buds, and premature
abscission (Rhoads et al. 1973). Similar symptoms are occasionally noted in
glasshouses, possibly as a result of ethylene accumulation from poorly ventilated burners.

Lily pollen responds differently to several air pollutants (Masaru et al.
1976). Pollen tube elongation is quite sensitive to SO_2 and NO_2 but not O_3
(Table 52). Combinations of SO_2 and NO_2, or O_3 and NO_2 result in greater
inhibition than the additive effects of the single gases. Pollen tube elongation is also sensitive to the aldehydes, formaldehyde and acrolein.

10.9 Orchid

The best known pollutant response of orchid (Cattleya) is its extreme
sensitivity to ethylene. Symptoms of ethylene injury have been described and
the disorder has been called "dried sepal injury" (Davidson 1949). The injury
is characterized by a progressive drying and bleaching of the sepals, beginning
at the tips and extending toward the bases, with maximum sensitivity at time
of bloom maturity. Very dilute concentrations of ethylene cause this injury
with 0.2 pphm ethylene for 24 hr or 10 pphm for eight hr resulting in sepal
drying.

Table 52. Pollen tube length (after 24 hr) of lily pollen after exposure to
air pollutants for 1, 2 or 5 hr at $28^{\circ}C$.

Gas	Concentration (pphm)	Exposure:	Ratio: A to B (%)[z] 1 hr	2 hr	5 hr
SO_2	40		96.2	92.4	0.0
	71		45.0	21.2	0.0
	250		32.2	0.0	0.0
NO_2	57		89.2	85.7	80.0
	170		77.5	60.3	17.2
	200		31.7	0.0	0.0
O_3	28		81.8	80.0	72.7
	209		88.2	95.5	80.0
Formaldehyde	37		100.0	100.0	27.7
H-CHO	140		86.5	67.3	0.0
	240		62.5	41.6	0.0
Acrolein	40		90.0	40.0	0.0
CH_2=CH-CHO	140		44.4	0.0	0.0
	170		0.0	0.0	0.0

[z] A: Pollen tube length after pollutant exposure.

B: Pollen tube length in clean air.

Data from Masaru et al. (1976)

10.10 Petunia

Petunia is a widely used bedding plant in temperate areas and many cultivars of differing colour and growth characteristics have been developed.
Petunia is rated as sensitive to both O_3 and PAN (Jacobson and Hill 1970) and
some cultivars are so sensitive that they could be used as indicator species
(Fig. 64). This species is more sensitive to either O_3 or SO_2 in terms of both
leaf injury and reduction in shoot and flower weight, than coleus, snapdragon,
marigold, celosia, impatiens or salvia (Adedipe et al. 1972a). Petunias respond
to O_3, PAN, SO_2 and other pollutants by developing chlorotic and necrotic markings on the foliage. Such injury is not only unsightly but may also indicate
reduced photosynthetic activity, growth and flower production. Injured leaves
on young plants also make the plant less saleable, therefore any procedure which
reduces pollutant injury will likely enhance crop value.

Oxidants. Petunia responses to oxidants have been thoroughly documented.
Oxidants suppress petunia growth (Taylor 1958). Growth rate is reduced to
about one-half by a three week exposure period compared with control plants
grown in clean air. Leaf size is also very much reduced and blossom initiation
completely suppressed by exposure to the reaction products of ozonated hexene.
Naturally occurring oxidants cause similar growth and blossom suppression and
induce typical oxidant injury on petunia leaves.

Fig. 64. Ozone-induced leaf flecking on petunia (Author's photo).

A large number of petunia cultivars have been separated into six classes of sensitivity to O_3 (Cathey and Heggestad 1972). Five of 65 cultivars are very insensitive and are considered to have the greatest potential in the development of new O_3 insensitive cultivars. Red cultivars are less sensitive than white. The use of petunia as an indicator species for oxidant has been evaluated using 'Snowstorm' petunia along with 'Bel-W3' tobacco to monitor O_3 effects in Georgia, U.S.A. (Walker and Barlow 1974). Petunia plants under antiozonant-treated cloth are generally taller and weigh more than those under non-treated cloth.

There are a number of physiological features involved in O_3-response of petunia. Analysis of petunia leaves for ascorbic acid reveals a parallel between O_3 insensitivity and ascorbic acid content, i.e. younger leaves have higher ascorbic acid content and O_3 insensitivity (Hanson et al. 1971). Cultivar differences in O_3 sensitivity are also reflected in ascorbic acid levels. Mature leaves increase in both ascorbic acid and O_3 insensitivity as the flowering stage is approached. Ozone-insensitive petunia hybrids are found to acquire a marked O_3 insensitivity upon visible differentiation of a flower bud (Hanson et al. 1975). Insensitive hybrids thus have some mechanism for protecting physiologically older leaves.

Oxidant affects petunia flowering (Feder 1970). Plants exposed to low levels of oxidant are slower to flower and form fewer flowers than those grown in filtered air. Flower size and weight decrease when plants are exposed to O_3 but removal from O_3 allows a return to normal flower development (Fig. 65). Decreased anthocyanin content of petunia flowers along with increased flower fresh weight, but not size, can be noted (Craker and Feder 1972).

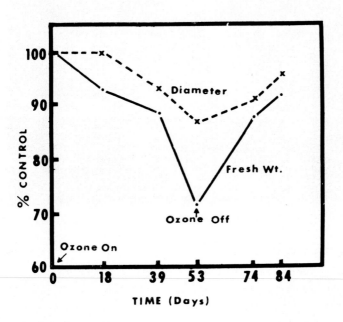

Fig. 65. Decline and recovery of flower wt and diam from O_3 stress. Flowers were harvested at the times indicated. Plants were exposed to 5-7 pphm O_3 for 6 hr per day, 5 days per week (Craker 1972).

Susceptibility of petunia to PAN has been included in studies of the inheritance of oxidant insensitivity (Hanson et al. 1976). Separate exposures to O_3, PAN and the ambient oxidant mixture in California, U.S.A., reveal wide differences among cultivars in sensitivity to each exposure regime. Petunias are more severely injured by PAN than by the O_3 component of the ambient oxidants.

Protection. Injury to petunia plants from O_3 or irradiated auto exhaust can be substantially reduced by withholding water from plants prior to pollutant exposure (Seidman et al. 1965). Petunia is protected from irradiated automobile exhaust by treatment with the growth retardant Phosphon D.

Ascorbic acid, an antioxidant, applied to petunias growing in oxidant-polluted air results in an increase in number and weight of leaves compared with untreated plants (Freebairn and Taylor 1960). Petunia is protected against O_3 injury by diphenylamine dusts (Gilbert et al. 1975).

Visible O_3-induced injury to petunias is reduced by application of growth regulators which retard elongation and promote dark green colouration (Cathey and Heggestad 1972). Among the chemicals most useful for this purpose are CBBP and SADH. Protection is increased by adding L-ascorbic acid and a wax coating to the spray solution. Some other growth retardants are ineffective in retarding either petunia growth or O_3 injury (Table 53).

Table 53. Ozone injury and ht of 'Pink Cascade' petunia treated with various chemicals.[x]

Chemical	Dosage (ppm)	Method of application[y]	O_3 injury rating[z]	Height at anthesis (cm)
Control (H_2O)	0	spray	8.0	20.5
SADH	5000	"	2.0	5.6
F724	"	"	2.5	5.9
C890	"	"	7.5	16.9
C011	"	"	2.5	6.1
F529	"	"	2.3	5.4
Ancymidol	10	drench	5.0	18.3
CBBP	100	"	3.5	9.4
Benomyl	100	"	8.5	21.3
Chlormequat	2000	"	6.5	18.9

[x] Spray or soil drench applied 39 days after seeding. Plants grown on long days and exposed to 45 pphm O_3 for 3 hr at anthesis.

[y] Sprayed until run-off or soil-drenched with 100 ml.

[z] Rating (on a 0-10 basis) after 48 hr.

Data from Cathey and Heggestad (1972)

Air Pollutant Comparison. A comparative study has been conducted by exposing 15 petunia cultivars to O_3, SO_2, PAN, NO_2 or irradiated automobile exhaust (Feder et al. 1969). Cultivars insensitive to one pollutant are often insensitive to others. The most sensitive cultivar is 'White Cascade' followed by 'Red Magic'. Cultivars are rated for sensitivity to each pollutant and an overall comparison is provided to allow cultivar selection for use in cities where all the pollutants may be present.

Other Pollutants. Injury to petunia in the field by a toxicant other than O_3 or SO_2 has been described (Rich and Tomlinson 1968). Symptoms include bands or spots of dull brown tissue and undersurface glazing; these have been called injury symptoms of "aldehyde", a pollutant which has not been fully character-ized.

10.11 Poinsettia

The wide range of air pollutant sensitivity in poinsettia indicates a potential for breeding and selection of insensitive cultivars in this species (Heggestad et al. 1973; Manning et al. 1973a). There is a range of cultivar sensitivity to either O_3 or SO_2 and those cultivars relatively insensitive to O_3 are also relatively insensitive to SO_2 (Cathey and Heggestad 1973). Expo-sure for three hr to 30 to 75 pphm O_3 or three hr to three ppm SO_2 results in visible injury to developing leaves.

The growth retardants ancymidol and chlormequat reduce visible injury on poinsettia plants exposed to O_3 or SO_2 (Cathey and Heggestad 1973). Chronic O_3 injury is also significantly reduced after benomyl soil drenches (Table 54).

Injury on poinsettia caused by 2,4-D includes spindly growth of stems and cupping of leaves (Fig. 66). Other herbicides may retard growth and injure leaves (Fig. 67).

10.12 Tulip

This species has been rated as sensitive to F and Cl_2 (Jacobson and Hill 1970). While F injury has been the principal concern of growers, a few studies with other air pollutants have been undertaken.

Sensitivity of tulip cultivars to SO_2 has been studied (Brennan et al. 1967). A wide range of injury among cultivars can be noted (Table 55).

Tulip injury apparently due to F pollution in the vicinity of industrial plants demonstrated wide differences in cultivar sensitivity (Spierings 1964). Exposure chambers are used to determine the effect of HF concentrations on leaf injury and tissue F levels. Tulips are among the most sensitive to F of monocotyledonous ornamental plants, with symptoms consisting of ivory-coloured necrosis of leaf tips (Spierings 1964; Wolting 1971).

In studies of illuminating gas toxicity to tulips and other bulb crops, the most significant phytotoxic constituent was found to be ethylene (Hitchcock et al. 1932). Plants exposed to illuminating gas have growth retardation with

Table 54. Effects of benomyl soil drenches on growth of poinsettia cultivars repeatedly exposed to a low level of O_3.

Plant responses	Air regimes[y]	Cultivars and benomyl drenches in μg/ml[x]					
		'Paul Mikkelsen'			'Mikkelwhite'		
		0	500	1000	0	500	1000
Chlorotic leaves	0	3.0d[z]	1.3e	0.6e	4.6d	1.6e	1.0e
	F	0	0	0.3e	0	0	0
Fallen leaves	0	3.0d	0.3e	0.6e	2.0d	0	0
	F	0	0.3e	0.6e	0	0.3e	0.3e
Stem diam (mm)	0	7.2d	8.3de	8.0de	7.3d	9.0e	8.6e
	F	9.0e	9.3e	9.3e	9.3e	9.0e	9.3e
Fresh wt tops (g)	0	17.6d	21.6ef	19.6e	17.0d	22.0e	22.6e
	F	23.3ef	24.6f	22.6ef	24.6e	24.3e	20.3e
Dry wt tops (g)	0	5.2d	8.3e	8.2e	5.8d	7.8e	7.0e
	F	9.2e	9.3e	8.6e	9.1e	10.3f	7.6e

[x] Benomyl drenches applied three times, 3, 30 and 57 days, after the initiation of the experiment.

[y] 0 = 10 pphm O_3, 8 hr/day, 5 days/wk, for 68 days; F = charcoal-filtered, O_3-free air.

[z] Mean separation within plant response by Duncan's multiple range test, 1% level.

Data from Manning et al. (1973b)

Table 55. Sulphur dioxide injury to tulip cultivars.

No injury		Slight	Medium	Severe
Czardas	Mrs. J. T. Scheers	Red Emperor	Queen of Night	Red Shine
Keiserkroon	Princess Margaret Rose	Garden Party	Magier	City of
Gudoshnick	White Triumphator	Rosy Wing	Gold Medal	Haarlem
Volcano	Prince of Austria	The Bishop	Praestens	Grand Hotel
Blizzard	L'Immortelle	Anjou	Fusilier	American
Texas Gold	Abraham Lincoln	Smiling Queen		Flag
Indian Chief	Igna Hume	Symphonia		
Cum Laude	White Emperor			
	George Grappe			

Data from Brennan et al. (1967)

death or abscission of leaves. Curling, looping, bending and inflation of leaves occurs depending on leaf age, growth rate and cultivar. Flower buds are killed at higher concentrations with the threshold dependent on treatment time and cultivar.

Fig. 66. Cupping of a poinsettia leaf exposed to 2,4-D in a commercial
glasshouse (Author's photo).

Fig. 67. Apparent herbicide injury on poinsettia (right) compared with normal
plant. The herbicide volatilized from spray applied under benches
before the crop was moved into the glasshouse (Author's photo).

10.13 Other Flower Crops

Fluoride. Crocus and ixias are very sensitive to HF along with freesia, gladiolus and tulips (Spierings 1969a). They have ivory-coloured necrosis of leaf tips. In contrast, narcissus and nerine have widespread yellow discolouration in the presence of HF.

Ethylene. Narcissus and hyacinth are sensitive to ethylene along with lily and tulip (Hitchcock et al. 1932). Growth retardation is observed with twisting of leaves followed by death or abscission. Leaf age, growth rate and cultivar affect leaf sensitivity, and treatment time and cultivar affect flower bud sensitivity.

African violet and marigold responses to ethylene have been reported (Abeles and Heggestad 1973). Exposure in growth chambers at concentrations comparable to those found in urban environments decreases flower number, as well as inhibiting leaf growth, promoting senescence and abscission, and inducing epinasty.

Herbicides. Injury to transplanted annual bedding plants by herbicides has been evaluated (Fretz 1976). Alachlor, a very effective herbicide for broadleaf weeds and grasses, injures salvia, and diphenamid, another effective

cides but do not injure bedding plants (Table 56).

Table 56. Post-transplant herbicide effects on visual toxicity on transplanted bedding plants.

	Visual phytotoxicity rating[z]				
	DCPA 11.2 kg/ha	Alachlor 3.4 kg/ha	Diphenamid 6.7 kg/ha	Trifluralin 2.2 kg/ha	Control
Ageratum	1.0	1.7	1.0	1.0	1.0
Amaranthus	4.3	2.3	2.7	4.0	1.7
Snapdragon	4.0	4.0	2.7	1.7	1.0
Celosia	2.3	1.7	5.7	1.0	1.0
Chrysanthemum	4.0	1.0	1.0	1.0	1.0
Dahlia	1.7	1.0	2.0	1.0	1.0
Pink	1.0	1.0	2.0	1.0	1.0
Geranium	1.0	1.0	1.0	1.0	1.0
Petunia 'Candy Apple'	1.7	1.0	1.0	1.0	1.0
Petunia 'Snow Cap'	1.0	1.0	1.0	1.0	1.0
Salvia	4.7	10.0	1.0	3.7	1.0
Marigold 'Lemon Drop'	1.0	1.3	1.0	1.0	1.0
Marigold 'Moonshot'	1.0	1.3	1.7	1.0	1.0
Verbena	1.0	1.0	1.0	1.0	1.0
Zinnia	1.7	1.7	1.0	1.0	1.0

[z] Visual rating scale: 1.0 (no injury) to 10.0 (complete plant kill).

Data from Fretz (1976)

CHAPTER 11

TURF GRASSES

Symptoms of air pollutant injury on turfgrasses include chlorophyll
destruction accompanied by spotting, yellowing and browning of leaf tissue.
The appearance as well as the growth of the turf is often adversely affected.
Air pollution may cause a decrease in growth rates of grass plants, particularly
ryegrass, without visible injury (Bleasdale 1952, 1973). There are marked
differences in air pollutant sensitivity among turfgrass species.

Oxidants. Annual bluegrass is widely distributed in lawns in temperate
regions and is extremely sensitive to oxidant smog (Bobrov 1955). The characteristic transverse chlorotic banding is related to leaf differentiation, cellular age and ability of stomata to open readily. The most sensitive cells are
those which have just completed maximum expansion.

Annual bluegrass has been suggested as a good indicator plant for oxidants
and attempts have been made to standardize growth of the species for this purpose (Juhren et al. 1957). The most sensitive plants are obtained by growing
them with 26° day and $20^{o}C$ night temperature, 16 hr photoperiod and natural
light. At warm temperature, sensitivity develops quickly, then diminishes to
insensitivity, while at cool temperature sensitivity develops later but lasts
much longer. The sensitivity changes if temperature is changed and injury is
correlated with the opening of stomata.

Oxidant smog injury to other turfgrasses has been reported (Hanson and
Juska 1969). Injury to 'Tifgreen' and 'Tifway' bermuda grass is observed in
California, U.S.A., and has led to the development of the smog-resistant bermuda
grass cultivar 'Santa Ana' for use in smog-prone areas.

Exposure of a large number of turfgrasses to severe industrial gaseous
pollution showed that quackgrass has the highest resistance followed by red
fescue and smooth bromegrass (Antipov 1959). Lawns exposed to air pollution
require more careful maintenance, and renovation of turf is required every two
or three years.

Several turfgrasses have been exposed to O_3 (Brennan and Halisky 1970).
Bentgrass and annual bluegrass are generally most sensitive to O_3-induced leaf
chlorosis (Table 57). Bermuda grass and zoysia are most insensitive. Cultivar
differences are noted, and are greatest at lower O_3 concentrations.

Table 57. Degree of injury and symptom expression in 11 turfgrasses exposed
to 30 pphm O_3 for 6 hr.

Turfgrass cultivar	Injury rating[y]	Type of foliage injury[z]
'Penncross' bentgrass	9	Tan necrosis, bleaching
Annual bluegrass	8	Tan to yellow necrosis, bleaching
'Merion' bluegrass	6	White necrosis, bleaching
'Lamora' ryegrass	6	Necrosis, dark-brown discolouration
'Manhattan' ryegrass	6	Necrosis, dark-brown discolouration
'Pennlawn' red fescue	6	Brown stippling, trace of bleaching
'Delta' bluegrass	4	White necrosis, bleaching
'P-16' bermuda grass	2	No necrosis, slight bleaching
'Highlight' red fescue	2	Some brown stippling
Common zoysia	0	None
'Meyer' zoysia	0	None

[y] Comparative degree of injury expressed numerically on a scale of 0 = no
injury to 9 = severe injury.

[z] Symptoms developed in grass foliage 7 days after O_3 exposure.

Data from Brennan and Halisky (1970)

Susceptibility of Kentucky bluegrass cultivars to O_3 in the field and in a
controlled exposure chamber has been determined (Wilton et al. 1972). Responses
are well correlated between field and chamber. 'Belturf' and 'Windsor' are the
most sensitive cultivars, and an insensitive strain has been identified. There
is also evidence that Kentucky bluegrass may develop some insensitivity to
repeated O_3 exposures.

Substantial reductions in O_3 injury can be obtained by benomyl or thiophanate applications to annual bluegrass as a soil amendment or drench (Moyer
et al. 1974a). Foliar sprays of benomyl are effective in reducing O_3 injury
to annual bluegrass, Kentucky bluegrass and creeping bentgrass but carbathiin
(carboxin) does not provide protection and itself causes leaf injury (Papple
and Ormrod 1977).

Sulphur Dioxide. There is ample evidence for SO_2-induced growth suppression in grasses (Bleasdale 1952, 1973). Exposure of turfgrasses to SO_2
indicates that red fescue and bentgrass are most sensitive to SO_2-induced leaf
necrosis (Table 58). Bermuda grass and zoysia are insensitive to SO_2 as they
are to O_3. There are also cultivar differences in SO_2 susceptibility. Sulphur
dioxide and O_3 sensitivity have been compared using Kentucky bluegrass cultivars
(Murray et al. 1975). Differential responses are generally obtained among
cultivars for both pollutants, and genetic relationships are indicated.

Table 58. Comparative injury sustained by 11 turfgrass cultivars after controlled 6-hr exposures to SO_2.

Cultivar	SO_2 concn (pphm) and injury rating[z]		
	75	85	180
'Highlight' red fescue	4	7	9
'Pennlawn' red fescue	2.5	7	9
'Penncross' bentgrass	1	7	9
Annual bluegrass	1	3	4
'Delta' bluegrass	0	3	4
'Merion' bluegrass	0	4	4
'Lamora' ryegrass	1	2	3
'Manhattan' ryegrass	0.5	0	2
'P-16' bermuda grass	0	0	0
Common zoysia	0	0	0
'Meyer' zoysia	0	0	0

[z] Comparative injury ratings expressed numerically on a scale of 0 = no injury to 9 = severe injury, scored 7 days after exposure.

Data from Brennan and Halisky (1970)

CHAPTER 12

WOODY PLANTS

There is a very extensive literature on air pollutant effects on forest
trees and only a few reviews will be noted here. Ornamental woody plants will
receive more detailed consideration. The nature and extent of air pollutant
injury to forest species has been reviewed with injury symptoms and sensitivity
ratings provided along with considerations of mimicking symptoms and meteorological effects (U.S. Department of Agriculture 1973). Information on susceptibility of woody plants to the oxidants O_3, PAN and NO_x, and to SO_2, has also
been compiled along with extensive lists of species sensitivities (Davis and
Wilhour 1976). It is noted that caution should be used in considering sensitivity ratings and that local environmental conditions and economic factors
should be emphasized for species of intermediate sensitivity. Lists of species
varying in sensitivity to O_3 and SO_2 have been prepared (Davis and Gerhold
1976). There are wide differences in susceptibility suggesting that arborists
can effectively choose trees for use in particular environments and plant
breeders can undertake improvement programmes leading to less sensitive hybrids
and cultivars.

12.1 Woody Ornamentals

Oxidants. The relative susceptibility of many landscape coniferous species
to O_3 has been determined (Davis and Wood 1972). Austrian and Scots pines are
susceptible while arborvitae and Colorado blue spruce are insensitive. Chlorotic mottle and tip necrosis are common injury symptoms but there is much variability among plants in both extent of injury and symptoms. The susceptibility
to O_3 of 15 species and/or cultivars of woody ornamentals has been reported
(Davis and Coppolino 1974). The most sensitive species are Hinodegiri Hirya
azalea, Korean azalea and tree-of-heaven (Table 59). Ozone injury is commonly
a tan or dark red to black stipple on the upper leaf surface.

Sugar maples are rated as relatively insensitive to O_3 but there is a
highly variable response among individuals (Hibben 1969a). The artificially-
induced injury symptoms are not found in urban and roadside environments
suggesting that O_3 is not an important causal factor of the sugar maple decline
or dieback disorder.

Nine woody species exposed to 30 pphm O_3 for an extended period demonstrated that there is a significant reduction by O_3 of height growth of

Table 59. Relative susceptibility and injury on plants exposed to 25 pphm
O_3 for 8 hr.

Common name	% Susceptible	Severity index[z]
Azalea, Hinodegiri	71.5	94.8
Azalea, Korean	81.0	70.0
Tree-of-Heaven	74.3	65.3
Chinese elm	80.0	24.5
Mock-orange, sweet	62.8	17.2
Viburnum, tea	17.8	5.4
Viburnum, linden	6.7	2.2
American holly, male	0	0
American holly, female	0	0
American linden	0	0
Amur privet	0	0
Black gum	0	0
Dense Anglojap yew	0	0
Hatfield Anglojap yew	0	0
Hetz Japanese holly	0	0
Mountain-laurel kalmia	0	0

[z] Based on [(severity factor) X (foliage injured) X (% population susceptible)] /100.

Data from Davis and Coppolino (1974)

sycamore, silver maple and sugar maple, and several species lose their leaves
sooner with O_3 treatment (Jensen 1973). Exposure of six woody species to 25
pphm O_3 for 110 days reduces height growth and leaf number and size of only
white birch seedlings (Jensen and Masters 1975).

Susceptibility of 13 woody shrubs and vines to O_3 has been rated (Davis
and Coppolino 1976). Staghorn sumac and Virginia creeper are sensitive and
white snowberry, bittersweet and Morrow honeysuckle are very insensitive
(Table 60). A dark pigmented stipple is the most common injury symptom and
leaves become less sensitive later in the growing season.

Variation in sensitivity to O_3 is found among red maple seedling progenies
from four seed sources with differences in the extent of injury, but not the
type of injury (Townsend and Dochinger 1974). Strong genetic control is indicated which should facilitate the selection of O_3-insensitive seedlings and
seed sources. Examination of the O_3 response of hybrid poplar (Populus
deltoides Bartr. X P. trichocarpa Torr. and Gray) cuttings indicates that
chronic O_3 exposure causes foliar injury and reduced growth (Jensen and
Dochinger 1974). Use of clonal repeatibility analyses on eastern white pine
demonstrates that the sensitivity-insensitivity reaction to combined O_3-SO_2
exposures is under strong genetic control (Houston and Stairs 1973).

Table 60. Relative susceptibility of woody shrubs and vines to 25 pphm O_3
for 8 hr.

	% Susceptible	Severity index[z]
Staghorn sumac	68.6	73.2
Virginia creeper	64.0	69.6
Indian currant (coral berry)	68.0	63.8
American elder (elderberry)	51.4	62.6
Dwarf ninebark	57.1	46.7
Multiflora rose	36.0	30.4
Smooth sumac	45.7	18.9
Red-osier dogwood	33.3	18.6
Silky dogwood	37.1	14.0
Autumn olive	17.1	12.3
White snowberry	11.4	0.8
Bittersweet	8.0	0.3
Morrow honeysuckle	3.3	0.0

[z] Based on [(severity factor) X (foliage injured) X (% population susceptible)] /100.

Data from Davis and Coppolino (1976)

The 'leaf roll necrosis' disorder of lilac has been the subject of
research projects and air pollutants have been implicated as the causal
agents. This foliar disorder is widespread in urban areas in the vicinity
of New York City, U.S.A. (Hibben and Walker 1966). The symptoms include leaf
roll, interveinal and marginal necrosis, chlorosis, bronzing and early leaf
abscission. (Fig. 68). This disorder is caused by air pollutants, including
oxidants and other unidentified pollutants (Hibben and Taylor 1974). There
has been evaluation of lilacs at several northeastern U.S.A. locations for
the leaf-roll necrosis disorder (Fig. 69, 70). A wide range of severity of
injury is found with lilac more sensitive than related species and inter-
specific hybrids. Insensitive cultivars are recommended for use in or near
large urban centres.

Benomyl drenches of from 60 to 100 mg/cm^3 of soil greatly reduce oxidant
injury on azaleas (Moyer et al. 1974b). Thiophanate foliar sprays at 500
and 750 mg/ml also reduce injury significantly.

Application of a complete (25-9-9 NPK) fertilizer to plantation-grown
white pines for Christman trees can alleviate air pollution-induced injury
symptoms and transform most dwarfed, thin and chlorotic trees into saleable
plants in one growing season (Will and Skelly 1974).

Fig. 68. Leaf roll necrosis symptoms on leaves of susceptible lilac at the
Brooklyn Botanic Garden, New York, U.S.A. (Photo courtesy of
C.R. Hibben).

Fig. 69. Lilac susceptible to leaf roll necrosis (cv. Katherine Havemeyer)
(Walker et al. 1975).

Fig. 70.   Lilac resistant to leaf roll necrosis (cv. Buffon) (Walker et al.
1975).

Sulphur Dioxide.   Four shade tree species exposed to SO_2 had the order
of decreasing sensitivity of Chinese elm, Norway maple, ginkgo and pin oak
(Table 61).   Long-term and short-term exposures do not affect the relative
susceptibility but in Norway maple and ginkgo SO_2 concentrations are more
important than time of exposure while in Chinese elm and pin oak the two
factors are of equal importance.   Exposure of hybrid poplars to SO_2 reveals
genetic differences in insensitivity to short-term high concentration treatments (Dochinger et al. 1972).   Foliar injury and subsequent shoot growth are
positively correlated.   Exposure of hybrid poplars to SO_2 at high concentrations for a short time or low rates for a much longer time both result in
foliar injury and shoot growth reduction (Dochinger and Jensen 1975).   Basal
and terminal cuttings respond differently to the long-term SO_2 exposure and
the two clones tested differ in sensitivity.   Examination of the responses of
a large number of woody plants to SO_2 or F showed that the majority of species
respond similarly to both pollutants (Dassler et al. 1972).

Table 61. Effects of acute exposures of urban trees to SO_2.

| | SO_2 ppm | % of leaf necrosis after exposure for: | | | |
		2 hr	4 hr	6 hr	8 hr
Chinese elm	2	10	50	70	90
	3	10	50	90	95
	4	10	90-100	90-100	90-100
Norway maple	2	0	0	<5	<10
	3	10	20	30	70
	4	10	30	60	80-100
Ginkgo	2	0	0	0	0
	3	0	20	20	60
	4	0-30	50	70	80
Pin oak	2	0	0	0	0
	3	0	0	0	0-5
	4	0	<1	0-5	0-10
	5	<1	0-10	0-20	0
	8	0	20	0-20	0-30

Data from Temple (1972)

Fluoride. Fifteen rose cultivars exposed to about two ppb HF for six
months varied widely in symptom expression with leaf injury ranging from mild
interveinal chlorosis to severe marginal necrosis (Brewer et al. 1967b). Top
dry weight is reduced in most cultivars but shoot number is increased by HF.
A greater number of flowers is produced but they are of poor quality.

Many ornamental shrub species have been exposed to short-term high concentrations of HF or NO_2 and classified according to sensitivity (MacLean et
al. 1968). Azalea, bougainvillea and pyracantha are very sensitive to both
while croton and carissa are relatively insensitive (Table 62). Some species
differ in relative sensitivity to the two gases, while others are moderately
sensitive to both.

Ethylene. Experiments on ethylene and illuminating gas sensitivity of
roses indicate that these pollutants cause epinasty of young leaves, abscission
of old and young leaves, yellowing along the veins, petal loss from cut
flowers, premature opening of buds, inhibition of shoot growth, and lateral
bud breaking (Zimmerman et al. 1931). In most cases, gas concentration, plant
age, temperature, time of exposure and cultivar are factors affecting response.

Trace Elements. The metal content of urban woody plants has been determined and only Fe, Pb, Na and Zn are present in above normal amounts, with Pb
contents generally higher in plants adjacent to primary highways or in areas
with natural Pb substrates (Smith 1973). White pine, red maple and Norway
spruce plants grown in a medium containing excessive Cd or Zn have reduced
root initiation, poor development of lateral roots, chlorosis, early leaf

Table 62. Relative susceptibility of 14 ornamental species to HF and NO_2.

	Very sensitive	Moderately sensitive	Relatively sensitive
HF	azalea	oleander	croton
	bougainvillea	Cape jasmine	carissa
	melaleuca	pittosporum	
	pyracantha	shore juniper	
	gardenia	hibiscus	
		ligustrum	
		ixora	
NO_2	azalea	pittosporum	carissa
	oleander	melaleuca	croton
	bougainvillea	ligustrum	shore juniper
	pyracantha	ixora	
	hibiscus	Cape jasmine	
		gardenia	

From MacLean et al. (1968). Reprinted with permission from Environ. Sci. Technol., 2, 444-9. Copyright by the American Chemical Society.

drop and other injury symptoms (Mitchell and Fretz 1977). White pine accumulates more Cd and Zn and is apparently less sensitive to Cd and Zn accumulation than the other two species.

Hemlock plants growing on old potato soils are injured by As accumulation (Sinclair et al. 1975) with the symptoms including scorch, premature casting of needles and death of fine roots. Hemlocks suffering from this As toxicity recover when moved to normal soils.

Salt. Soil salts. Almost all ornamental plants are considered to be sensitive to salt (Hayward and Bernstein 1958), but there are some species which are moderately salt insensitive and which could be used in saline soil situations. Ornamental shrubs are important for landscaping salt-affected areas so salt sensitivity of this group should receive special attention. The use of artificially salinized soil plots and sand cultures to study salt tolerance of 25 shrub and ground cover species showed insensitive species to include bougainvillea, Natal plum and rosemary, and sensitive species, star jasmine, guava, holly and rose (Bernstein et al. 1972). Twenty species of trees and shrubs exposed to different salt ($NaCl + CaCl_2$ 1:1) levels in soil illustrated that black locust, honey locust, Russian olive, squaw bush and tamarix survive high salt levels while blue spruce, Douglas fir, black walnut, linden, barberry, euonymous, multiflora rose, spiraea and golden willow have little salt insensitivity (Monk and Peterson 1962).

Azalea and gardenia are particularly sensitive to salinity (Lunt et al. 1957). Symptoms of salt injury in azaleas include leaf tips turning reddish-brown and older leaves becoming necrotic and abscising (Lunin and Stewart 1961). In camellias, the injury consists of necrosis of tips of older leaves followed by abscission. There are cultivar differences in salinity sensitivity of azalea (Lunt et al. 1957).

Salt water. Roses are sensitive to irrigation with water containing even moderate levels of chlorides. Experiments with 'Baccara' roses on Rosa chinensis rootstock irrigated with saline water show that water containing chlorides is more harmful than water containing nitrates at the same salinity level (Yaron et al. 1969). Stem and leaf growth and water uptake decrease with increasing soil salinity. There is a cumulative effect of soil salinity even at low levels.

Azaleas are relatively sensitive to nutrient imbalance in irrigation water according to Lunt et al. (1956). High bicarbonate is particularly undesirable because it increases the supply of Na when Ca is low, and decreases the absorption of Ca and Fe. The effects of bicarbonate also may be due to carbonic acid, CO_2 in the soil atmosphere, or pH changes.

Salt spray. An extensive list of roadside tree and shrub sensitivity to aerial salt spray has been prepared (Table 63). There are very great differences in sensitivity and many species are very insensitive to salt spray. Such species can be readily used in problem areas, such as along roadways on which deicing salt is used. The insensitivity of many species may also account for apparent disparities in research reports. For example, there is little injury to trees after exposure to many applications of NaCl or $CaCl_2$ even though many species have more tissue Cl in leaves and twigs than control plants (Holmes 1961). Oaks are particularly insensitive to salt application.

Study of the maple decline disorder has included both salt spray and soil salt effects. The disorder is confined to roadside and road drainage areas (Lacasse and Rich 1964). Elevated Na and soluble salt levels are noted in maple tissue from these areas. Neither fungi nor nematodes are a satisfactory alternate explanation to salt for the decline disorder. Studies of the injury on sugar maple leaves near roadways have attributed it to salt accumulation (Hall et al. 1973). The leaves on the roadway side of trees have higher concentrations of Na and Cl than those on the side away from the roadway. Soil Na and conductivity are also higher on the roadway side of trees. There is no difference in essential element composition between the two sides of the trees.

Another aspect of salt spray injury relates to the use of saline water in cooling towers at electric generation stations. Studies of spray drift from such installations have been reported (McCune et al. 1977). Stage of development, species and individual phenotype are all factors in determining plant

Table 63.   Sensitivity of roadside trees and shrubs to aerial drift of deicing
           salt.

Common name	Rating[z]	Common name	Rating	Common name	Rating
Deciduous trees		Lombardy poplar	2-3	Japanese barberry	2
		Pear	2-3	Burningbush	2
Norway maple	1	Basswood	2-3	Forsythia	2-3
Horse-chestnut	1	Crabapple	3	Privet	2-3
Tree-of-heaven	1	Largetooth aspen	3	Alder buckthorn	2-3
Cottonwood	1	Trembling aspen	3	Speckled alder	3
Black locust	1	Apple	3-4	Flowering quince	3-4
Sugar maple	1-2	Bur oak	3-4	Gray dogwood	3-4
Shagbark Hickory	1-2	Hawthorn	4	Beauty-bush	3-4
Honey locust	1-2	Manitoba maple	4-5	Bumalda spiraea	3-4
Black walnut	1-2	Allegheny service		Red Osier dogwood	4-5
English walnut	1-2	berry	4-5		
Choke cherry	1-2	White mulberry	4-5	Conifers	
Red oak	1-2	Beech	5		
Silver maple	2			Blue spruce	1
White ash	2	Deciduous shrubs		Jack pine	1-2
Poplar	2			Mugo pine	1-2
Black willow	2	Siberian pea-tree	1	Tamarack	2
Mountain ash	2	European buckthorn	1	Austrian pine	2
White elm	2	Honeysuckle	1-2	Juniper	2-3
Chinese elm	2	Staghorn sumac	1-2	Norway spruce	3
Red maple	2-3	Japanese lilac	1-2	White cedar	3-4
White birch	2-3	Common lilac	1-2	Yew	4
Grey birch	2-3	Russian olive	1-3	White spruce	4-5
Catalpa	2-3	Mock orange	1-3	Scots pine	4-5
Quince	2-3	European cran-		Hemlock	4-5
		berry bush	1-3	White pine	5

[z] Rating of 1 = no injury. Rating of 5 represents complete dieback and needle
browning of conifers, and complete dieback, evidence of previous injury and
lack of flowering of deciduous species. Ratings of 2, 3 and 4 encompass
slight, moderate and extensive gradations of these symptoms.
Data from Lumis et al (1973)

susceptibility. The youngest leaves are most sensitive in deciduous woody
species but year-old conifer needles are most sensitive. The order of decreasing sensitivity is Canadian hemlock, white ash, white flowering dogwood,
forsythia, chestnut oak, silk tree, black locust, red maple, eastern white
pine, golden rain tree and witch hazel.

CHAPTER 13

THE FUTURE OF POLLUTION AND HORTICULTURE

Pollution will continue to impinge on horticultural production everywhere and may increase in intensity as a result of increased industrialization and fossil fuel burning together with accumulation of trace elements and pesticides. There is a need for environmental quality standards for all types of pollution in relation to horticultural production to provide a basis for abatement of pollution at its sources. In addition, optimal combinations of horticultural species and cultivars with particular management practices will go a long way toward lessening the impact of pollution. This is particularly true for horticultural plants which exhibit much species and cultivar diversity in response to pollution. This diversity makes it very difficult to assess fully the impact of pollutants on fruit, vegetable and ornamental plant production and the protection of these plants from pollution injury. The distinctive behaviour of species and, in many cases, cultivars means that important pollutant problems must be studied separately for each strain with a distinctive response pattern.

Air pollution research needs have been thoroughly reviewed (Heck et al. 1973). An extensive list of suggested areas of research is provided and it is clear that much remains to be done. A chart of interrelationships to indicate the diversity of research needs has been developed (Fig. 71). Effects of pollutant exposure at any stage of the life cycle and under the range of environments usually encountered during plant growth must be evaluated together with the genetic bases for variation in pollutant response.

Studies of the biochemical, physiological and morphological relationships of horticultural plant sensitivity are incomplete. Most biochemical studies to date have involved cellular organelles, especially mitochondria and chloroplasts. Fewer studies have been concerned with membrane permeability, isozymes, hormone levels or translocation, yet all these approaches could be valuable. Studies involving metabolism should parallel whole plant studies to secure a greater understanding of the pollution problem as well as to uncover possible means of alleviating injury to plants. Elimination of the many gaps in our present knowledge should lead ultimately to an accurate assessment of air pollution effects on the range of significant horticultural species, and then to workable injury control procedures.

Recommendations for research on the relationship between air pollution and trees (Smith and Dochinger 1975) can be readily adapted for pollution studies

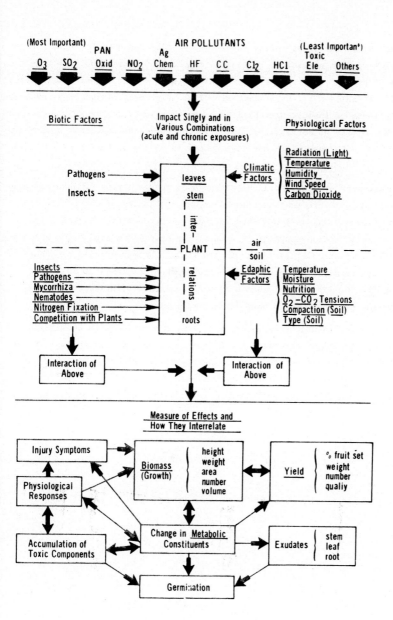

Fig. 71. Research needs in the area of air pollution effects on vegetation. These conceptual interrelations may be viewed from a cellular, whole plant, plant species, or plant community level. Thus, an effect on any of the listed processes could reduce an organism's competitive advantage (Heck et al. 1973).

in horticulture. The top priority research objective would be to obtain dose-response information on visible responses to pollutants with experiments on the influence of genetic factors, environmental factors, and pollutant interactions. Next would be analyses of the ability of pollutant stress to predispose or aggravate the effects of insect, microbial or biotic stresses, followed by determination of the ability of horticultural plants to reduce environmental contamination, and the usefulness of insensitive cultivars in the reduction of pollution stress.

High research priority would be assigned to the gathering of dose-response information on invisible responses including an evaluation of the ability of pollution stress to reduce growth and influence reproduction. Another high priority would be the development of accurate, simple and reproducible methodology to identify and quantify injury in the field. Research would also be conducted to determine the physiological and biochemical bases of pollution stress, and to develop reliable and economically sound cultural procedures for protecting horticultural plants against pollution.

The needs for future research are thus many and varied with enough requirements and opportunities to command the attention of many groups for many years. Horticultural plants are not only very susceptible to pollution stresses but also very responsive to amelioration practices. Thus, research on pollution in horticulture is likely to be rewarding to both the industry and the investigators.

REFERENCES

ABELES, F.B. and HEGGESTAD, H.E. (1973). Ethylene: an urban air pollutant. J. Air Pollut. Contr. Ass., 23, 517-21. 18,183

ADAMS, D.F. (1961). An air pollution phytotron: A controlled environment facility for studies into the effects of air pollutants on vegetation. J. Air Pollut. Contr. Ass., 11 470-6. 55

ADAMS, D.F. and SULZBACH, C.W. (1961). Nitrogen deficiency and fluoride susceptibility of bean seedlings. Science, 133, 1425-6. 40,41,121

ADAMS, D.F., HENDRIX, J.W. and APPLEGATE, H.G. (1957). Relationship among exposure periods, foliar burn, and fluorine content of plants exposed to hydrogen fluoride. J. Agr. Food Chem., 5, 108-16. 16,17,33

ADEDIPE, N.O. and ORMROD, D.P. (1972). Hormonal regulation of ozone phytotoxicity in Raphanus sativus. Z. Pflanzenphysiol., 68, 254-8. 45,46,137

ADEDIPE, N.O. and ORMROD, D.P. (1974). Ozone-induced growth suppression in radish plants in relation to pre- and post-fumigation temperatures. Z. Pflanzenphysiol., 71, 281-7. 35,36,137

ADEDIPE, N.O., BARRETT, R.E. and ORMROD, D.P. (1972a). Phytotoxicity and growth responses of ornamental bedding plants to ozone and sulfur dioxide. J. Am. Soc. Hort. Sci., 97, 341-5. 164,176

ADEDIPE, N.O., HOFSTRA, G. and ORMROD, D.P. (1972b). Effects of sulfur nutrition on phytotoxicity and growth responses of bean plants to ozone. Can. J. Bot., 50, 1789-93. 35,40,118

ADEDIPE, N.O., KHATAMIAN, H. and ORMROD, D.P. (1973). Stomatal regulation of ozone phytotoxicity in tomato. Z. Pflanzenphysiol., 68, 323-8. 30,33,37,43,143

AKESSON, W.B. and YATES, W.E. (1964). Problems relating to application of agricultural chemicals and resulting drift residues. Annu. Rev. Entom., 9, 285-318. 81

ALLMENDINGER, D.G., MILLER, V.L. and JOHNSON, F. (1950). The control of fluorine scorch of gladiolus with foliar dust and sprays. Proc. Am. Soc. Hort. Sci., 56, 427-32. 41,47,175

AMER, S.M. and ALI, E.M. (1968). Cytological effects of pesticides II. Meiotic effects of some phenols. Cytologia, 33, 21-33. 111

AMER, S.M. and FARAH, O.R. (1968). Cytological effects of pesticides III. Meiotic effects of N-methyl-1-naphthyl carbamate "Sevin". Cytologia, 33, 337-44. 111

ANDERSON, L.D. and ATKINS, E.L., JR. (1968). Pesticide usage in relation to beekeeping. Ann. Rev. Entom., 13, 213-38. 80

ANTIPOV, U.G. (1960). Influence of smoke and gas on the flowering and fruiting of some trees and shrubs. Sbornik. Botanischekikh Rabof., 2, 167-72.   184

APPLEGATE, H.G. and ADAMS, D.F. (1960a). Nutritional and water effect on fluoride uptake and respiration of bean seedlings. Phyton, 14, 111-20.
                                                                                  40,121

APPLEGATE, H.G. and ADAMS, D.F. (1960b). Effect of atmospheric fluoride on respiration of bush beans. Bot. Gaz., 121, 223-7.   107

APPLEGATE, H.G. and ADAMS, D.F. (1960c). "Invisible injury" of bush beans by atmospheric and aqueous fluorides. Int. J. Air Wat. Pollut., 3, 231-48.   107

ARGAUER, R.J., MASON, H.C., CORLEY, C., HIGGINS, A.H., SAULS, J.N. and LILJEDAHL, L.A. (1968). Drift of water-diluted and undiluted formulations of malathion and azinphosmethyl applied by airplane. J. Econ. Ent., 61, 1015-20.   82

ARNDT, U. (1974). The Kautsky-effect: a method for the investigation of the actions of air pollutants in chloroplasts. Environ. Pollut., 6, 182-194.   104,105

ARVIK, J.H. and ZIMDAHL, R.L. (1974a). Barriers to the foliar uptake of lead. J. Environ. Qual., 3, 369-73.   72

ARVIK, J.H. and ZIMDAHL, R.L. (1974b). The influence of temperature, pH, and metabolic inhibitors on uptake of lead by plant roots. J. Environ. Qual., 3, 374-6.   72

AUERMANN, E., JACOBI, J., ETERNACH, R. and KUHN, H. (1976). Untersuchungen über den bleigehalt pflanzlicher nahrungsmittel im wirkungsbereich eines bleiemitherenden betriebes. Die Nahrung, 20, 509-18.   71

AYERS, J.C., JR. and BARDEN, J.A. (1975). Net photosynthesis and dark respiration of apple leaves as affected by pesticides. J. Am. Soc. Hort. Sci., 100, 24-8.   108,154

BALLANTYNE, D.J. (1973). Sulphite inhibition of ATP formation in plant mitochondria. Phytochemistry, 12, 1207-9.   106

BALLANTYNE, D.J. (1977). Sulphite oxidation by mitochondria from green and etiolated peas. Phytochemistry, 16, 49-50.   106

BARDEN, L.E. and HANAN, J.J. (1972). Effects of ethylene on carnation keeping life. J. Am. Soc. Hort. Sci., 97, 785-8.   166

BARKER, A.V. and CRAKER, L.E. (1973). Effects of soil and water pollution on agriculture. In: Manning, W.J., Ed. The Impact of Environmental Stresses on Agriculture, pp. 26-39, Publ. 89. Tech. Guidance Centre for Environ. Qual., U. of Mass., Amherst, Mass.   91

BARKER, W.G. (1972). Toxicity levels of mercury, lead, copper, and zinc in tissue culture systems of cauliflower, lettuce, potato and carrot. Can. J. Bot., 50, 973-6.   75,147

BARRETT, T.W. and BENEDICT, H.W. (1970) Sulfur dioxide. In: Jacobson, J.S. and Hill, A.C., Eds., Recognition of Air Pollution Injury to Vegetation: A Pictorial Atlas, pp. C1-C17, Air Pollut. Contr. Ass., Pittsburgh, Pa.   13

BARTON, L.V. (1940). Toxicity of ammonia, chlorine, hydrogen cyanide, hydrogen sulphide, and sulphur dioxide gases. IV. Seeds. Contrib. Boyce Thompson Inst., 11, 357-63.   19,137

BAUMHARDT, G.R. and WELCH, L.F. (1972). Lead uptake and corn growth with soil-applied lead. J. Environ. Qual., 1, 92-4.   71,126

BAZZAZ, F.A., CARLSON, R.W. and ROLFE, G.L. (1974a). The effect of heavy metal on plants: part I. Inhibition of gas exchange in sunflower by Pb, Cd, Ni and Tl. Environ. Pollut., 7, 241-6.   105

BAZZAZ, F.A., CARLSON, R.W. and ROLFE, G.L. (1975). Inhibition of corn and sunflower photosynthesis by lead. Physiol. Pl., 34, 326-9.   72

BAZZAZ, F.A., ROLFE, G.L. and CARLSON, R.W. (1974b). Effect of Cd on photosynthesis and transpiration of excised leaves of corn and sunflower. Physiol. Pl., 32, 373-6.   69,105

BAZZAZ, F.A., ROLFE, G.L. and WINDLE, P. (1974c). Differing sensitivity of corn and soybean photosynthesis and transpiration to lead contamination. J. Environ. Qual. 3, 156-8.   105

BAZZAZ, M.B. and GOVINDJEE. (1974). Effects of cadmium nitrate on spectral characteristics and light reactions of chloroplasts. Environ. Lett., 6, 1-12.   105

BEAVINGTON, F. (1975). Heavy metal contamination of vegetables and soil in domestic gardens around a smelting complex. Environ. Pollut., 9, 211-7. 75

BENEDICT, H.M., MILLER, C.J. and OLSON, R.E. (1971). Economic impact of air pollutants on plants in the United States. Stanford Res. Inst., Menlo Park, Calif. 77 pp.   7

BENEDICT, H.M., ROSS, J.M. and WADE, R.W. (1964). The disposition of atmospheric fluorides by vegetation. Int. J. Air Wat. Pollut., 8, 279-89.
                                                     16,129,139,146

BENNETT, J.H. and HILL, A.C. (1973). Inhibition of apparent photosynthesis by air pollutants. J. Environ. Qual., 2, 526-30.   104

BENNETT, J.H. and HILL, A.C. (1974). Acute inhibition of apparent photosynthesis by phytotoxic air pollutants. In: Dugger, M., Ed. Air Pollution Effects on Plant Growth, ACS Symp. Ser. 3, pp. 115-127, Am. Chem. Soc., Washington, D.C.   102

BENNETT, J.H. and HILL, A.C. (1975). Interactions of air pollutants with canopies of vegetation. In: Mudd, J.B. and Kozlowski, T.T., Eds., Responses of Plants to Air Pollution, pp. 273-306, Academic Press, Inc., New York.   29

BENNETT, J.H., HILL, A.C., SOLEIMANI, A. and EDWARDS, W.H. (1975). Acute effects of combinations of sulphur dioxide and nitrogen dioxide on plants. Environ. Pollut. 9, 127-32.   24,25

BENNETT, J.P. and OSHIMA, R.J. (1976). Carrot injury and yield response to ozone. J. Am. Soc. Hort. Sci., 101, 638-9.   124

BENNETT, J.P., RESH, H.M. and RUNECKLES, V.C. (1974). Apparent stimulations of plant growth by air pollutants. Can. J. Bot., 52, 35-41.   47

202

BENSON, N.R.   (1959).   Fluoride injury or soft suture and splitting of
     peaches.   Proc. Am. Soc. Hort. Sci., 74, 184-98.   17,41,159

BENSON, N.R.   (1976).   Retardation of apple tree growth by soil arsenic
     residues.   J. Am. Soc. Hort. Sci., 101, 251-3.   66

BERNSTEIN, L. and HAYWARD, H.E.   (1958).   Physiology of salt tolerance.
     Annu. Rev. Pl. Physiol., 9, 25-46.   93

BERNSTEIN, L., FRANCOIS, L.E. and CLARK, R.A.   (1972).   Salt tolerance of
     ornamental shrubs and ground covers.   J. Am. Soc. Hort. Sci., 97, 550-6.
                                                                              95,193

BIDWELL, R.G.S. and FRASER, D.E.   (1972).   Carbon monoxide uptake and meta-
     bolism by leaves.   Can. J. Bot., 50, 1435-9.   19

BISCOE, P.V., UNSWORTH, M.H. and PINCKNEY, H.R.   (1973).   The effects of low
     concentrations of sulphur dioxide on stomatal behaviour in Vicia faba.
     New Phytol., 72, 1299-306.   30

BISHOP, R.F. and CHISHOLM, D.   (1962).   Arsenic accumulation in Annapolis
     Valley orchard soils.   Can. J. Soil Sci., 42, 77-80.   60

BLEASDALE, J.K.A.   (1952).   Atmospheric pollution and plant growth.   Nature,
     169, 376-7.   184,185

BLEASDALE, J.K.A.   (1973).   Effects of coal smoke pollution gases on the
     growth of ryegrass (Lolium perenne L.).   Environ. Pollut., 5, 275-85.
                                                                              184,185

BOBROV, R.A.   (1955).   The leaf structure of Poa annua with observations of
     its smog sensitivity in Los Angeles County.   Am. J. Bot., 42, 467-74.   184

BOLTON, J.   (1975).   Liming effects on the toxicity to perennial ryegrass of
     a sewage sludge contaminated with zinc, nickel, copper and chromium.
     Environ. Pollut., 9, 295-304.   77

BONTE, J. and LONGUET, P.   (1975).   Interrelations entre la pollution par le
     dioxyde de soufre et le mouvement des stomates chez le Pelargonium x
     hortorum: effets de l'humidite relative et de la teneur en gaz
     carbonique de l'air.   Physiol. Veg., 13, 527-37.   30,173

BOVAY, E.   (1969).   Effets de l'anhydride sulfureux et des composes fluores
     sur la vegetation.   In: Air Pollution.   Proc. 1st Eur. Congr. Influence
     of Air Pollution on Plants and Animals, 1968, pp. 111-35, Pudoc,
     Wageningen.   13,16

BRANDT, C.C. and HECK, W.W.   (1968).   Effects of air pollutants on vegetation.
     In: Stern, A.C., Ed., Air Pollution.   Vol. 1, pp. 401-43, Academic Press
     Inc., New York.   4

BRASHER, E.P., FIELDHOUSE, D.J. and SASSER, M.   (1973).   Ozone injury in
     potato variety trials.   Pl. Dis. Rep., 57, 542-4.   133

BREIDENBACH, A.W.   (1965).   Pesticide residues in air and water.   Archs.
     Environ. Hlth., 10, 827-30.   82,84

BRENNAN, E. and HALISKY, P.M.   (1970).   Response of turfgrass cultivars to
     ozone and sulfur dioxide in the atmosphere.   Phytopathology, 60, 1544-6.
                                                                              184,185,186

BRENNAN, E. and LEONE, I.A. (1968). The response of plants to sulfur dioxide or ozone-polluted air supplied at varying flow rates. Phytopathology 58, 1661-4.   41,127,143

BRENNAN, E. and LEONE, I.A. (1969). Air pollution damage to chrysanthemum foliage. Pl. Dis. Rep., 53, 54-5.   168

BRENNAN, E. and LEONE, I.A. (1972). Chrysanthemum response to sulfur dioxide and ozone. Pl. Dis. Rep., 56, 85-7.   168

BRENNAN, E. and RHOADS, A. (1976). Response of field-grown bean cultivars to atmospheric oxidant in New Jersey. Pl. Dis. Rep., 60, 941-5.   115

BRENNAN, E.G., LEONE, I.A. and DAINES, R.H. (1950). Fluorine toxicity in tomato as modified by alterations in the nitrogen, calcium, and phosphorus nutrition of the plant. Pl. Physiol., 25, 736-47.   40,145

BRENNAN, E., LEONE, I. and DAINES, R.H. (1962). Ammonia injury to apples and peaches in storage. Pl. Dis. Rep., 46, 792-5.   19,30,150,159

BRENNAN, E.G., LEONE, I.A. and DAINES, R.H. (1964). The importance of variety in ozone plant damage. Pl. Dis. Rep., 48, 923-4.   133,134

BRENNAN, E., LEONE, I.A. and DAINES, R.H. (1965). Chlorine as a phytotoxic air pollutant. Int. J. Air Wat. Pollut., 9, 791-7.   19

BRENNAN, E., LEONE, I.A. and DAINES, R.H. (1967). Characterization of the plant damage problem by air pollutants in New Jersey. Pl. Dis. Rep., 51, 850-4.   5,180,181

BRENNAN, E., LEONE, I.A. and DAINES, R.H. (1970). Toxicity of sulfur dioxide to highbush blueberry. Pl. Dis. Rep., 54, 704-6.   159

BRENNAN, E., LEONE, I.A. and HOLMES, C. (1969). Accidental chlorine gas damage to vegetation. Pl. Dis. Rep., 53, 873-5.   19

BREWER, R.F., GUILLEMET, F.B. and CREVELING, R.K. (1962). Influence of N-P-K fertilization on incidence and severity of oxidant injury to mangels and spinach. Soil Sci., 92, 298-301.   39,138

BREWER, R.F., GUILLEMET, F.B. and SUTHERLAND, F.H. (1966). The effect of atmospheric fluoride on gladiolus growth, flowering, and corm production. Proc. Am. Soc. Hort. Sci., 88, 631-4.   174

BREWER, R.F., McCOLLOCH, R.C. and SUTHERLAND, F.H. (1957). Fluoride accumulation in foliage and fruit of wine grapes growing in the vicinity of heavy industry. Proc. Am. Soc. Hort. Sci., 70, 183-8.   161

BREWER, R.F., SUTHERLAND, F.H. and PEREZ, R.O. (1969a). The effects of simulated rain and dew on fluoride accumulation by citrus foliage. J. Am. Soc. Hort. Sci., 94, 284-6.   41,158

BREWER, R.F., SUTHERLAND, F.H. and GUILLEMET, F.B. (1969b). Application of calcium sprays for the protection of citrus from atmospheric fluorides. J. Am. Soc. Hort. Sci., 94, 302-4.   47,159,175

BREWER, R.F., SUTHERLAND, F.H. and GUILLEMET, F.B. (1969c). Effects of various fluoride sources on citrus growth and fruit production. Environ. Sci. Technol., 3, 378-81.   158

BREWER, R.F., CREVELING, R.K., GUILLEMET, F.B. and SUTHERLAND, F.H. (1960a).
The effects of hydrogen fluoride gas on seven citrus varieties. Proc.
Am. Soc. Hort. Sci., 75, 236-43.   158

BREWER, R.F., GARBER, M.J., GUILLEMET, F.B. and SUTHERLAND, F.H. (1967a).
The effects of accumulated fluoride on yields and fruit quality of
'Washington' navel oranges. Proc. Am. Soc. Hort. Sci., 91, 150-6.   158

BREWER, R.F., SUTHERLAND, F.H. and GUILLEMET, F.B. (1967b). The relative
susceptibility of some popular varieties of roses to fluoride air pollution. Proc. Am. Soc. Hort. Sci., 91, 771-6.   192

BREWER, R.F., SUTHERLAND, F.H., GUILLEMET, F.B. and CREVELING, R.K. (1960b).
Some effects of hydrogen fluoride gas on bearing navel orange trees.
Proc. Am. Soc. Hort. Sci., 76, 208-14.   158

BULL, J.N. and MANSFIELD, T.A. (1974). Photosynthesis in leaves exposed
to SO_2 and NO_2. Nature, 250, 443-4.   104

BURTON, K.W. and JOHN, E. (1977). A study of heavy metal contamination in
the Rhondda Fawr, South Wales. Wat., Air, Soil Pollut., 7, 45-68.   74

CAMERON, J.W. (1975). Inheritance in sweet corn for resistance to acute
ozone injury. J. Am. Soc. Hort. Sci., 100, 577-9.   113

CAMERON, J.W. and TAYLOR, O.C. (1973). Injury to sweet corn inbreds and
hybrids by air pollutants in the field and by ozone treatments in the
greenhouse. J. Environ. Qual., 2, 387-9.   125

CAMERON, J.W., JOHNSON, H., JR., TAYLOR, O.C. and OTTO, H.W. (1970).
Differential susceptibility of sweet corn hybrids to field injury
by air pollution. HortScience, 5, 217-9.   125

CANNON, H.L. and BOWLES, J.M. (1962). Contamination of vegetation by
tetraethyl lead. Science, 137, 765-6.   71

CANTWELL, A.M. (1968). Effect of temperature on response of plants to ozone
as conducted in a specially designed plant fumigation chamber. Pl.
Dis. Rep., 52, 957-60.   55,56

CAPRON, T.M. and MANSFIELD, T.A. (1975). Generation of nitrogen oxide
pollutants during CO_2 enrichment of glasshouse atmospheres. J. Hort.
Sci., 50, 233-8.   13

CAPRON, T.M. and MANSFIELD, T.A. (1976). Inhibition of net photosynthesis
in tomato in air polluted with NO and NO_2. J. Exp. Bot., 27, 1181-6.
                                                                       103,145

CARLSON, R.W. and BAZZAZ, F.A. (1977). Growth reduction in American
sycamore (Plantanus occidentalis L.) caused by Pb-Cd interaction.
Environ. Pollut., 12, 243-53.   75

CARLSON, R.W., BAZZAZ, F.A. and ROLFE, G.L. (1975). The effect of heavy
metals on plants II. Net photosynthesis and transpiration of whole
corn and sunflower plants treated with Pb, Cd, Ni, and Tl. Environ.
Res., 10, 113-20.   104

CARNEY, A.W., STEPHENSON, G.R., ORMROD, D.P. and ASHTON, G.C. (1973).
Ozone-herbicide interactions in crop plants. Weed Sci., 21, 508-11. 26,143

CARO, J.H. and TAYLOR, A.W. (1971). Pathways of loss of dieldrin from soils under field conditions. J. Agr. Food Chem., 19, 379-84.   83

CATHEY, H.M. and HEGGESTAD, H.E. (1972). Reduction of ozone damage to Petunia hybrida Vilm. by use of growth regulating chemicals and tolerant cultivars. J. Am. Soc. Hort. Sci., 97, 695-700.   177,179

CATHEY, H.M. and HEGGESTAD, H.E. (1973). Effects of growth retardants and fumigations with ozone and sulfur dioxide on growth and flowering of Euphorbia pulcherrima Willd. J. Am. Soc. Hort. Sci., 98, 3-7.   45,180

CHAMEL, A. and GARREC, J.P. (1977). Penetration of fluorine through isolated pear leaf cuticles. Environ. Pollut., 12, 307-10.   17

CHANG, C.W. (1970). Effect of fluoride on ribosomes and ribonuclease from corn roots. Can. J. Biochem., 48, 450-4.   108

CHANG, C.W. (1973). Biochemical and biophysical investigation into growth and aging of corn seedlings treated with fluoride. Fluoride, 6, 162-78.   108

CHANG, C.W. (1975). Fluorides. In: Mudd, J.B. and Kozlowski, T.T., Eds., Responses of Plants to Air Pollution, pp. 57-95, Academic Press, Inc., New York.   16

CHANG, C.W. and HEGGESTAD, H.E. (1974). Effect of ozone on photosystem II in Spinacea oleracea chloroplasts. Phytochemistry, 13, 871-3.   103

CHANG, C.W. and THOMPSON, C.R. (1965). Subcellular distribution of fluoride in navel orange leaves. Int. J. Air Wat. Pollut., 9, 685-91.   17

CHANG, C.W. and THOMPSON, C.R. (1966). Site of fluoride accumulation in navel orange leaves. Pl. Physiol., 41, 211-3.   111

CHISHOLM, D. (1972). Lead, arsenic, and copper content of crops grown on lead arsenate-treated and untreated soils. Can. J. Pl. Sci., 52, 583-8. 72

CHISHOLM, D. and SPECHT, H.B. (1967). Effect of application rates of disulfoton and phorate, and of irrigation on aphid control and residues in canning peas. Can. J. Pl. Sci., 47, 175-80.   84

CIRELI, B. (1976). Observations on the effects of some air pollutants on Zea mays leaf tissue. Phytopath. Z., 86, 233-9.   111

CLAYBERG, C.D. (1971). Screening tomatoes for ozone resistance. HortScience, 6, 396-7.   141

COSTONIS, A.C. (1973). Injury to eastern white pine by sulfur dioxide and ozone alone and in mixtures. Eur. J. Forest Pathol., 3, 50-5.   24

COTTENIE, A., DHAESE, A. and CAMERLYNCK, R. (1976). Plant quality response to uptake of polluting elements. Qual. Plantarum, 26, 293-319.   65

COULSON, C. and HEATH, R.L. (1974). Inhibition of the photosynthetic capacity of isolated chloroplasts by ozone. Pl. Physiol., 53, 32-8.   102

COULSON, C. and HEATH, R.L. (1975). The interaction of peroxyacetyl nitrate (PAN) with the electron flow of isolated chloroplasts. Atmos. Environ., 9, 231-8.   103

CRAKER, L.E. (1971). Ethylene production from ozone injured plants.
Environ. Pollut., 1, 299-304.   106

CRAKER, L.E. (1972). Decline and recovery of petunia flower development
from ozone stress. HortScience, 7, 484.   178

CRAKER, L.E. and FEDER, W.A. (1972). Development of the inflorescence in
petunia, geranium, and poinsettia under ozone stress. HortScience, 7,
59-60.   178

CRAKER, L.E. and STARBUCK, J.S. (1972). Metabolic changes associated with
ozone injury of bean leaves. Can. J. Pl. Sci., 52, 589-97.   105

CROCKER, W. and KNIGHT, L.I. (1908). Effect of illuminating gas and ethylene upon flowering carnations. Bot. Gaz., 46, 259-76.   165

CUNNINGHAM, J.D., KEENEY, D.R. and RYAN, J.A. (1975a). Yield and metal
composition of corn and rye grown on sewage sludge-amended soil.
J. Environ. Qual., 4, 448-54.   77,126

CUNNINGHAM, J.D., RYAN, J.A. and KEENEY, D.R. (1975b). Phytotoxicity in
and metal uptake from soil treated with metal-amended sewage sludge.
J. Environ. Qual., 4, 455-60.   77,126

CUNNINGHAM, J.D., KEENEY, D.R. and RYAN, J.A. (1975c). Phytotoxicity and
uptake of metals added to soils as inorganic salts or in sewage sludge.
J. Environ. Qual., 4, 460-2.   77

CURTIS, C.R. and HOWELL, R.K. (1971). Increases in peroxidase isoenzyme
activity in bean leaves exposed to low doses of ozone. Phytopathology,
61, 1306-7.   101

CURTIS, L.R., EDGINGTON, L.V. and LITTLEJOHNS, D.J. (1975). Oxathiin
chemicals for control of bronzing in white beans. Can. J. Pl. Sci.,
55, 151-6.   120

CZUBA, M. and ORMROD, D.P. (1974). Effects of cadmium and zinc on ozone-
induced phytotoxicity in cress and lettuce. Can. J. Bot., 52, 645-9.
                                                                          26,66,126

DAINES, R.H. (1969a). Air pollution and plant response. In: Gunckel,
J.E., Ed., Current Topics in Plant Science, pp. 436-53, Academic
Press, New York.   4

DAINES, R.H. (1969b). Sulfur dioxide and plant response. In: Air Quality
Monographs No. 69-8, American Petroleum Institute, New York, U.S.A.   13

DAINES, R.H., LEONE, I.A. and BRENNAN, E. (1960a). Air pollution as it
affects agriculture in New Jersey. New Jersey Agr. Exp. Sta. Bull.
794, 1-14.   5

DAINES, R.H., LEONE, I.A., BRENNAN, E. and MIDDLETON, J.T. (1960b). Damage
to spinach and other vegetation in New Jersey from ozone and other air-
borne oxidants. Phytopathology, 50, 570 (Abs.).   138

DARLEY, E. (1966). Studies on the effect of cement-kiln dust on vegetation.
J. Air Pollut. Contr. Ass., 16, 145-50.   20

DARLEY, E.F. and MIDDLETON, J.T. (1966). Problems of air pollution in
plant pathology. Annu. Rev. Phytopath., 4, 103-18.   35

DARLEY, E.F., NICHOLS, C.W. and MIDDLETON, J.T. (1966). Identification of air pollution damage to agricultural crops. The Bulletin, Calif. Dept. Agr., 55, No. 1, 19 pp.   4

DARLEY, E.F., MIDDLETON, J.T. and KENDRICK, J.B., JR. (1956). Sulfur dioxide injury on citrus. Calif. Agr., 10(1), 9.   157

DASS, H.C. and WEAVER, G.M. (1968). Modification of ozone damage to Phaseolus vulgaris by antioxidants, thiols and sulfhydryl reagents. Can. J. Pl. Sci., 48, 569-74.   42,101,118

DASS, H.C. and WEAVER, G.M. (1972). Enzymatic changes in intact leaves of Phaseolus vulgaris following ozone fumigation. Atmos. Environ., 6, 759-63.   101

DASSLER, H.-G. VON, RANFT, H. and REHN, K.-H. (1972). Zur Widerstands- fahigkeit von Geholzen gegenuber Fluorverbindungen und Schwefeldioxid. Flora, 161, 289-302.   191

DAVIDSON, O.W. (1949). Effects of ethylene on orchid flowers. Proc. Am. Soc. Hort. Sci., 53, 440-6.   175

DAVIS, D.D. and COPPOLINO, J.B. (1974). Relative ozone susceptibility of selected woody ornamentals. HortScience, 9, 537-9.   187,188

DAVIS, D.D. and COPPOLINO, J.B. (1976). Ozone susceptibility of selected woody shrubs and vines. Pl. Dis. Rep., 60, 876-8.   188,189

DAVIS, D. and GERHOLD, H.D. (1976). Selection of Trees for Tolerance of Air Pollutants. In: Better Trees for Metropolitan Landscapes Symp. Proc., pp. 61-6, USDA Forest Service. Gen. Tech. Rep. NE-22.   187

DAVIS, D.D. and KRESS, L. (1974). The relative susceptibility of ten bean varieties to ozone. Pl. Dis. Rep., 58, 14-16.   115

DAVIS, D.D. and SMITH, S.H. (1974). Reduction of ozone-sensitivity of Pinto bean by bean common mosaic virus. Phytopathology, 64, 383-5.   28

DAVIS, D.D. and SMITH, S.H. (1975). Bean common mosaic virus reduces ozone sensitivity of pinto bean. Environ. Pollut., 9, 97-101.   28

DAVIS, D.D. and SMITH, S.H. (1976). Reduction of ozone sensitivity of Pinto bean by virus-induced local lesions. Pl. Dis. Rep., 60, 31-4.   28

DAVIS, D.D. and WILHOUR, R.G. (1976). Susceptibility of woody plants to sulfur dioxide and photochemical oxidants. A literature review. Ecological Research Series EPA-600/3-76-102. U.S. Env. Prot. Agency, Corvallis, Oregon, U.S.A.   187

DAVIS, D.D. and WOOD, F.A. (1972). The relative susceptibility of eighteen coniferous species to ozone. Phytopathology, 62, 14-9.   187

DAY, B.E. (1970). Horticulture and pollution. HortScience, 5, 237-9. Preface

DEDOLPH, R., TER HAAR, G., HOLTZMAN, R. and LUCAS, H., JR. (1970). Sources of lead in perennial ryegrass and radishes. Environ. Sci. Technol., 4, 217-23.   71,137

DE MAURA, J., LE TOURNEAU, D. and WIESE, A.C. (1973). The response of potato (Solanum tuberosum) tuber phosphoglucomutase to fluoride. Biochem. Physiol. Pflanzen, 164, 228-33.   108

DE ONG, E.R.   (1946).   Injury to apricot leaves from fluorine deposit.
Phytopathology, 36, 469-71.   155

DIRR, M.A.   (1975).   Effects of salts and application methods on English
ivy.   HortScience, 10, 182-4.   96

DOCHINGER, L.S.   (1972).   Can trees cleanse the air of particulate pollutants?   Proc. Int. Shade Tree Conf., 48, 45-8.   31

DOCHINGER, L.S. and JENSEN, K.F.   (1975).   Effects of chronic and acute
exposure to sulphur dioxide on the growth of hybrid poplar cuttings.
Environ. Pollut., 9, 219-29.   191

DOCHINGER, L.S., TOWNSEND, A.M., SEEGRIST, D.W. and BENDER, F.W.   (1972).
Responses of hybrid poplar trees to sulfur dioxide fumigation.
J. Air Pollut. Contr. Ass., 22, 369-71.   191

DONOHO, C.W., JR.   (1964).   Influence of pesticide chemicals on fruit set,
return bloom, yield, and fruit size of the apple.   Proc. Am. Soc. Hort.
Sci. 85, 53-9.   81,154

DUGGER, M., Ed.   (1974).   Air Pollution Effects on Plant Growth.   ACS Symp.
Ser. 3, Am. Chem. Soc., Washington, D.C., 150 pp.   100

DUGGER, W.M. and TING, I.P.   (1970).   Air pollution oxidants -- their
effects on metabolic processes in plants.   Annu. Rev. Pl. Physiol.,
21, 215-34.   100

DUGGER, W.M., JR., KOUKOL, J. and PALMER, R.L.   (1966).   Physiological and
biochemical effects of atmospheric oxidants on plants.   J. Air Pollut.
Contr. Ass., 16, 467-71.   100

DUGGER, W.M., JR., TAYLOR, O.C., CARDIFF, E. and THOMPSON, C.R.   (1962).
Stomatal action in plants as related to damage from photochemical
oxidants.   Pl. Physiol., 37, 487-91.   30

DUNNING, J.A. and HECK, W.W.   (1973).   Response of Pinto bean and tobacco
to ozone as conditioned by light intensity and/or humidity.   Environ.
Sci. Technol., 7, 824-6.   34,118

DUNNING, J.A. and HECK, W.W.   (1977).   Response of bean and tobacco to
ozone: effect of light intensity, temperature and relative humidity.
J. Air Pollut. Contr. Ass., 27, 882-6.   34,118

DUNNING, J.A., HECK, W.W., TINGEY, D.T.   (1974).   Foliar sensitivity of
pinto bean and soybean to ozone as affected by temperature, potassium
nutrition and ozone dose.   Wat., Air, Soil Pollut., 3, 305-13.   118

DVORAK, J.   (1968).   Influence of simazine on shoot and root growth of
apples and plums.   Weed Res., 8, 8-13.   90,155

EDWARDS, C.   (1970).   Persistent Pesticides in the Environment.   CRC Press,
pp. 1-78.   84

EDWARDS, J.H., HORTON, B.D. and KIRKPATRICK, H.C.   (1976).   Aluminum toxicity
symptoms in peach seedlings.   J. Am. Soc. Hort. Sci., 101, 139-42.   65

EHLIG, C.F. and BERNSTEIN, L.   (1959).   Foliar absorption of sodium and
chloride as a factor in sprinkler irrigation.   Proc. Am. Soc. Hort.
Sci., 74, 661-70.   95

ELFVING, D.C., GILBERT, M.D., EDGERTON, L.J., WILDE, M.H. and LISK, D.J.
(1976). Antioxidant and antitranspirant protection of apple foliage
against ozone injury. Bull. Environ. Contam. Toxicol., 15, 336-41.   150

ENGLE, R.L. and GABELMAN, W.H. (1966). Inheritance and mechanism for
resistance to ozone damage in onion, Allium cepa L. Proc. Am. Soc.
Hort. Sci., 89, 423-30.   30,113,114

ENGLE, R.L. and GABELMAN, W.H. (1967). The effects of low levels of ozone
on Pinto beans, Phaseolus vulgaris L. Proc. Am. Soc. Hort. Sci., 91,
304-9.   47

ENGLE, R.L., GABELMAN, W.H. and ROMANOWSKI, R.R., JR. (1965). Tipburn,
an ozone incited response in onion, Allium cepa L. Proc. Am. Soc.
Hort. Sci. 86, 468-74.   129

ESHEL, Y. and KATAN, J. (1972a). Effects of day and night temperature on
tolerance of solanaceous vegetables to diphenamid. HortScience, 7,
67-68.   85

ESHEL, Y. and KATAN, J. (1972b). Effect of dinitroanilines on solanaceous
vegetables and soil fungi. Weed Sci., 20, 243-6.   85

ESHEL, Y. and KATAN, J. (1972c). Effect of time of application of diphenamid on pepper, weeds, and disease. Weed Sci., 20, 468-71.   84,85

EVANS, L.S. and TING, I.P. (1973). Ozone-induced membrane permeability
changes. Am. J. Bot., 60, 155-62.   100

FACTEAU, T.J. and ROWE, R.E. (1977). Effect of hydrogen fluoride and
hydrogen chloride on pollen tube growth and sodium fluoride on pollen
germination in 'Tilton' apricot. J. Am. Soc. Hort. Sci., 102, 95-6.   155

FACTEAU, T.J., WANG, S.Y. and ROWE, K.E. (1973). The effect of hydrogen
fluoride on pollen germination and pollen tube growth in Prunus avium
L. cv. 'Royal Ann'. J. Am. Soc. Hort. Sci., 98, 234-6.   17,155

FAIRCHILD, E.J., MURPHY, S.D. and STOKINGER, H.E. (1959). Protection by
sulfur compounds against the air pollutants ozone and sulfur dioxide.
Science, 130, 861-2.   40

FASSETT, D.W. (1975). Cadmium: biological effects and occurrence in the
environment. Annu. Rev. Pharmacol., 15, 425-35.   66

FAVRETTO, L., MARLETTA, G.P. and GABRIELLI, L.F. (1975). Pollution of
vineyards by atmospheric lead. J. Sci. Food Agr., 26, 987-92.   161

FEDER, W.A. (1970). Plant response to chronic exposure of low levels of
oxidant type air pollution. Environ. Pollut., 1, 73-9.   165,173,178

FEDER, W.A. (1973). Cumulative effects of chronic exposure of plants to
low levels of air pollutants. In: Naegele, J.A., Ed., Air Pollution
Damage to Vegetation, Adv. Chem. Ser. 122, pp. 21-30, Am. Chem. Soc.,
Washington, D.C.   5

FEDER, W.A. and CAMPBELL, F.J. (1968). Influence of low levels of ozone
on flowering of carnations. Phytopathology, 58, 1038-9.   165

FEDER, W.A., CAMPBELL, F.J. and BUTTERFIELD, N.W. (1967). Tip burn, an
oxidant incited response of carnations. Pl. Dis. Rep., 51, 793.   165

FEDER, W.A., FOX, F.L., HECK, W.W. and CAMPBELL, F.J. (1969). Varietal responses of petunia to several air pollutants. Pl. Dis. Rep., 53, 506-10.
52,179

FETNER, R.H. (1958). Chromosome breakage in Vicia faba by ozone. Nature, 181, 504-5.
109

FERENBAUGH, R.W. (1976). Effects of simulated acid rain on Phaseolus vulgaris L. (Fabaceae). Am. J. Bot., 63, 283-8.
21,122

FINLAYSON, D.G., WILLIAMS, I.H., BROWN, M.J. and CAMPBELL, C.J. (1976). Distribution of insecticide residues in carrots at harvest. J. Agr. Food Chem., 24, 606-8.
124

FISHERIES AND ENVIRONMENT CANADA (1976). Criteria for national air quality objectives. Sulphur dioxide, suspended particulates, carbon monoxide, oxidants (ozone) and nitrogen dioxide. Ottawa, Canada. 41 pp.

FLETCHER, R.A., ADEDIPE, N.O. and ORMROD, D.P. (1972). Abscisic acid protects bean leaves from ozone-induced phytotoxicity. Can. J. Bot., 50, 2389-91.
45,47,119

FLOOR, H. and POSTHUMUS, A.C. (1977). Biologische erfassung von ozone - und PAN-immissionen in den Niederlanden 1973, 1974 und 1975. VDI - Berichte 270, 183-90.
51

FLORE, J.A. and BUKOVAC, M.J. (1976). Pesticide effects on the plant cuticle: II. EPTC effects on leaf cuticle morphology and composition in Brassica oleracea L. J. Am. Soc. Hort. Sci., 101, 586-90.
90

FRANK, R., ISHIDA, K. and SUDA, P. (1976a). Metals in agricultural soils of Ontario. Can. J. Soil Sci., 56, 181-96.
83

FRANK, R., BRAUN, H.E., ISHIDA, K. and SUDA, P. (1976b). Persistent organic and inorganic pesticide residues in orchard soils and vineyards of southern Ontario. Can. J. Soil Sci., 56, 463-84.
82,153

FREEBAIRN, H.T. (1963). Uptake and movement of $1-C^{14}$ ascorbic acid in plants. Physiol. Pl., 16, 517-22.
42,118

FREEBAIRN, H.T. and TAYLOR, O.C. (1960). Prevention of plant damage from air-borne oxidizing agents. Proc. Am. Soc. Hort. Sci., 76, 693-9.
42,118,179

FRETZ, T.A. (1976). Herbicide performance on transplanted annual bedding plants. HortScience, 11, 110-1.
183

FRICK, H. and CHERRY, J.H. (1974). An early site of physiological damage to soybean and cucumber seedlings following ozonation. In: Dugger, M., Ed., Air Pollution Effects on Plant Growth, ACS Symp. Ser. 3, pp. 128-47, Am. Chem. Soc., Washington, D.C.
126

FUJIWARA, T. (1970). Sensitivity of grapevines to injury by atmospheric sulfur dioxide. J. Jap. Soc. Hort. Sci., 39, 219-23.
161,162

GABELMAN, W.H. (1970). Alleviating the effects of pollution by modifying the plant. HortScience, 5, 250-2.
112

GARDNER, J.O. and ORMROD, D.P. (1976). Response of the Rieger begonia to ozone and sulphur dioxide. Scientia Hort., 5, 171-81.
165

GENTILE, A.G., FEDER, W.A., YOUNG, R.E. and SANTNER, Z. (1971). Susceptibility of Lycopersicon spp. to ozone injury. J. Am. Soc. Hort. Sci.,
96, 94-6.   110,140,141

GILBERT, M.D., MAYLIN, G.A., ELFVING, D.C., EDGERTON, L.J., GUTENMANN, W.H.
and LISK, D.J. (1975). The use of diphenylamine to protect plants
against ozone injury. HortScience, 10, 228-31.   44,120,129,150,179

GODZIK, S. and SASSEN, M.M.A. (1974). Einwirkung von SO$_2$ auf die
Feinstruktur der Chloroplasten von Phaseolus vulgaris. Phytopath. Z.,
79, 155-9.   111

GOLDSMITH, C.D., JR., SCANLON, P.F. and PIRIE, W.R. (1976). Lead concentrations in soil and vegetation associated with highways of different
traffic densities. Bull. Environ. Contam. Toxicol., 16, 66-70.   71

GROVES, A.B. and ROLLINS, H.A., JR. (1966). A study of the comparative
influence of captan and dodine fungicides on fruit set, fruit size,
return bloom and cropping of the Stayman apple. HortScience, 1, 11-12. 154

GUDERIAN, R. (1969). Obstbau in gebieten mit schwefeldioxid-immissionen.
Erwerbsobstbau 11, 110-3.   159

GUDERIAN, R. (1977). Air Pollution. Phytotoxicity of Acidic Gases and
its Significance in Air Pollution Control. Ecol. Stud. 22. Springer-
Verlag, Berlin, 127 pp.   4

GUTTAY, A.J.R. (1976). Impact of deicing salts upon the endomycorrhizae
of roadside sugar maples. Soil Sci. Soc. Am. J., 40, 952-4.   97

HAAGEN-SMIT, A.J., DARLEY, E.F., ZAITLIN, M., HULL, H. and NOBLE, W. (1952).
Investigation on injury to plants from air pollution in the Los Angeles
area. Pl. Physiol., 27, 18-34.   9

HAAS, J.H. (1970). Relation of crop maturity and physiology to air pollution incited bronzing of Phaseolus vulgaris. Phytopathology, 60, 407-10.
                                                                         114

HAGHIRI, F. (1974). Plant uptake of cadmium as influenced by cation
exchange capacity, organic matter, zinc, and soil temperature.
J. Environ. Quality, 3, 180-3.   68

HAISMAN, D.R. (1974). The effect of sulphur dioxide on oxidizing enzyme
systems in plant tissues. J. Sci. Food Agr., 25, 803-10.   101

HALEVY, A.H. and KOFRANEK, A.M. (1977). Silver treatment of carnation
flowers for reducing ethylene damage and extending longevity.
J. Am. Soc. Hort. Sci., 102, 76-7.   168

HALL, R., HOFSTRA, G. and LUMIS, G.P. (1973). Leaf necrosis of roadside
sugar maple in Ontario in relation to elemental composition of soil
and leaves. Phytopathology, 63, 1426-7.   97,194

HALLGREN, J.-E. and HUSS, K. (1975). Effects of SO$_2$ on photosynthesis and
nitrogen fixation. Physiol. Pl., 34, 171-6.   107

HANAN, J.J. (1973a). Ethylene pollution from combustion in greenhouses.
HortScience, 8, 23-4.   18

HANAN, J.J. (1973b). Ethylene dosages in Denver and marketability of cut-
flower carnations. J. Air Pollut. Contr. Ass., 23, 522-4.   166,167

HAND, D.W. (1972). Air pollution in glasshouses and the effects of aerial pollutants on crops. Scientia Hort., 24, 142-52.   6

HANSON, A.A. and JUSKA, F.V. (1969). Turfgrass Science, Am. Soc. Agron., Madison, Wisc.   184

HANSON, G.P., ADDIS, D.H. and THORNE, L. (1976). Inheritance of photo-chemical air pollution tolerance in petunias. Can. J. Gen. Cyt., 18, 579-92.   113,178

HANSON, G.P., THORNE, L. and ADDIS, D.H. (1975). The ozone sensitivity of Petunia hybrida Vilm. as related to physiological age. J. Am. Soc. Hort. Sci., 100, 188-90.   178

HANSON, G.P. THORNE, L. and JATIVA, C.D. (1971). Ozone tolerance of petunia leaves as related to their ascorbic acid concentration. In: Englund, H.M. and Beery, W.T., Eds., Proc. 2nd Internat. Clean Air Congr., pp. 261-6, Academic Press, New York.   106,177

HARWARD, M. and TRESHOW, M. (1975). Impact of ozone on the growth and reproduction of understory plants in the aspen zone of western U.S.A. Environ. Conserv., 2, 17-23.   10

HASEK, R.F., JAMES, H.A. and SCIARONI, R.H. (1969). Ethylene - its effect on flower crops. Florists Rev., 144: 3721, 21, 65-68, 79-82; 3722, 16-17, 53-56.   18

HASSETT, J.J., MILLER, J.E. and KOEPPE, D.E. (1976). Interaction of lead and cadmium on maize root growth and uptake of lead and cadmium by roots. Environ. Pollut., 11, 297-302.   126

HAYWARD, H.E. and BERNSTEIN, L. (1958). Plant-growth relationships on salt-affected soils. Bot. Rev., 24, 584-635.   94,123,193

HEAGLE, A.S. (1973). Interactions between air pollutants and plant para-sites. Annu. Rev. Phytopath., 11, 365-88.   26

HEAGLE, A.S., BODY, D.E. and POUNDS, E.K. (1972). Effect of ozone on yield of sweet corn. Phytopathology, 62, 683-7.   60,125

HEAGLE, A.S., BODY, D.E. and HECK, W.W. (1973). An open-top field chamber to assess the impact of air pollution on plants. J. Environ. Qual., 2, 365-8.   61

HEAGLE, A.S., HECK, W.W. and BODY, D. (1971). Ozone injury to plants as influenced by air velocity during exposure. Phytopathology, 61, 1209-12.   41

HEATH, R.L. (1975). Ozone. In: Mudd, J.B. and Kozlowski, T.T., Eds., Responses of Plants to Air Pollution, pp. 23-55, Academic Press, Inc. New York.   11

HEATH, R.L., CHIMIKLIS, P and FREDERICK, P. (1974). The role of potassium and lipids in ozone injury to plant membranes. In: Dugger, M., Ed., Air Pollution Effects on Plant Growth, Amer. Chem. Soc. Symp. Ser. 3, pp. 58-75, Am. Chem. Soc., Washington, D.C.   100

HECK, W.W. (1964). Plant injury induced by photochemical reaction products of propylene-nitrogen dioxide mixtures. J. Air Pollut. Contr. Ass., 14, 255-61.   9,18

HECK, W.W. (1966). The use of plants as indicators of air pollution.
Int. J. Air Wat. Pollut., 10, 99-111.   50

HECK, W.W. (1968). Factors influencing expression of oxidant damage to
plants. Annu. Rev. Phytopath., 6, 165-88.   31

HECK, W.W. (1973). Air pollution and the future of agricultural production. In: Naegele, J.A., Ed., Air Pollution Damage to Vegetation,
Adv. Chem. Ser. 122, pp. 118-129, Am. Chem. Soc., Washington, D.C.   6

HECK, W.W., DAINES, R.H. and HINDAWI, I.J. (1970). Other phytotoxic
pollutants. In: Jacobson, J.S. and Hill, A.C., Eds., Recognition
of Air Pollution Injury to Vegetation: A Pictorial Atlas, pp. F1 -
F24, Air Pollut. Contr. Ass., Pittsburg, Pa.   19

HECK, W.W., DUNNING, J.A. and HINDAWI, I.J. (1965). Interactions of
environmental factors on the sensitivity of plants to air pollution.
J. Air Pollut. Contr. Ass., 15, 511-5.   31,35,37

HECK, W.W., DUNNING, J.A. and JOHNSON, H. (1968). Design of a simple plant
exposure chamber. Nat. Center Air Pollut. Contr. Publ. APTD-68-6.   64

HECK, W.W., TAYLOR, O.C. and HEGGESTAD, H.E. (1973). Air pollution
research needs: herbaceous and ornamental plants and agriculturally
generated pollutants. J. Air Pollut. Contr. Ass., 23, 257-66.   197,198

HEGGESTAD, H.E. (1968). Diseases of crops and ornamental plants incited
by air pollutants. Phytopathology, 58, 1089-97.   4

HEGGESTAD, H.E. (1973). Photochemical air pollution injury to potatoes in
the Atlantic coastal states. Am. Potato J., 50, 315-28.   133

HEGGESTAD, H.E. (1974). Air pollutants from, and effects on, agriculture.
In: Air Pollut. Contr. Ass., Ed., Control Technology for Agricultural
Air Pollutants, pp. 170-177, Pittsburgh, Pa.   4

HEGGESTAD, H.E. and HECK, W.W. (1971). Nature, extent, and variation of
plant response to air pollutants. Adv. Agron., 23, 111-45.   4

HEGGESTAD, H.E., TUTHILL, K.L. and STEWART, R.N. (1973). Differences among
poinsettias in tolerance to sulfur dioxide. HortScience, 8, 337-8.   180

HENDRIX, J.W. and HALL, H.R. (1958). The relationship of certain leaf
characteristics and flower color to atmospheric fluoride-sensitivity
in gladiolus. Proc. Am. Soc. Hort. Sci., 72, 503-10.   174

HEPTING, G.H. (1968). Diseases of forest and tree crops caused by air
pollutants. Phytopathology, 58, 1098-101.   4

HEXTER, A.C. and GOLDSMITH, J.R. (1971). Carbon monoxide: association of
community air pollution with mortality. Science, 172, 265-7.   19

HEWITT, E.J. (1953). Metal interrelationships in plant nutrition. I.
Effects of some metal toxicities on sugar beet, tomato, oat, potato,
and marrowstem kale grown in sand culture. J. Exp. Bot., 4, 59-64.   75

HIBBEN, C.R. (1969a). Ozone toxicity in sugar maple. Phytopathology,
50, 1423-8.   187

HIBBEN, C.R. (1969b). The distinction between injury to tree leaves by
ozone and mesophyll-feeding leaf hoppers. Forest Sci., 15, 154-7.   28

HIBBEN, C.R. and TAYLOR, M.P. (1974). The leaf roll-necrosis disorder of lilacs: etiological role of urban-generated air pollutants. J. Am. Soc. Hort. Sci., 99, 508-14.   189

HIBBEN, C.R. and WALKER, J.T. (1966). A leaf roll-necrosis complex of lilacs in an urban environment. Proc. Am. Soc. Hort. Sci., 89, 636-42. 189

HILL, A.C. (1967). A special purpose plant environmental chamber for air pollution studies. J. Air Pollut. Contr. Ass., 17, 743-8.   54,55

HILL, A.C. (1969). Air quality standards for fluoride vegetation effects. J. Air Pollut. Contr. Ass., 19, 331-6.   18

HILL, A.C. (1971). Vegetation: a sink for atmospheric pollutants. J. Air Pollut. Contr. Ass., 21, 341-6.   29

HILL, A.C. and LITTLEFIELD, N. (1969). Ozone. Effect on apparent photosynthesis, rate of transpiration, and stomatal closure in plants. Environ. Sci. Technol., 3, 52-6.   102,103

HILL, A.C., HEGGESTAD, H.E. and LINZON, S.N. (1970). Ozone. In: Jacobson, J.S. and Hill, A.C., Eds., Recognition of Air Pollution Injury to Vegetation: A Pictorial Atlas, pp. B1-B22, Air Pollut. Contr. Ass., Pittsburg, Pa.   10

HILL, A.C., TRANSTRUM, L.G., PACK, M.R. and HOLLOMAN, A., JR. (1959). Facilities and techniques for maintaining a controlled fluoride environment in vegetation studies. J. Air Pollut. Contr. Ass., 9, 22-7.   59

HILL, A.C., TRANSTRUM, L.G., PACK, M.R. and WINTERS, W.S. (1958). Air pollution with relation to agronomic crops: VI. An investigation of the "hidden injury" theory of fluoride damage to plants. Agron. J., 50, 562-5.   145

HILL, A.C., PACK, M.R., TRESHOW, M., DOWNS, R.J. and TRANSTRUM, L.G. (1961). Plant injury induced by ozone. Phytopathology, 51, 356-63.   10,123

HINDAWI, I.J. (1968). Injury by sulfur dioxide, hydrogen fluoride, and chlorine as observed and reflected on vegetation in the field. J. Air Pollut. Contr. Ass., 18, 307-12.   5

HINESLY, T.D., JONES, R.L., ZIEGLER, E.L. and TYLER, J.J. (1977). Effects of annual and accumulative applications of sewage sludge on assimilation of zinc and cadmium by corn (Zea mays L.). Environ. Sci. Technol., 11, 182-8.   77

HITCHCOCK, A.E., CROCKER, W. and ZIMMERMAN, P.W. (1932). Effect of illuminating gas on the lily, narcissus, tulip, and hyacinth. Contrib. Boyce Thompson Inst., 4, 155-76.   180,183

HITCHCOCK, A.E., ZIMMERMAN, P.W. and COE, R.R. (1962). Results of ten years' work (1951-1960) on the effect of fluorides on gladiolus. Contrib. Boyce Thompson Inst., 21, 303-44.   174

HODGSON, R.H. and HOFFER, B.L. (1977a). Diphenamid metabolism in pepper and an ozone effect. I. Absorption, translocation, and the extent of metabolism. Weed Sci., 25, 324-30.   26

HODGSON, R.H. and HOFFER, B.L. (1977b). Diphenamid metabolism in pepper and an ozone effect. II. Herbicide metabolite characterization. Weed Sci., 25, 331-7.   26

HODGSON, R.H., DUSBABEK, K.E. and HOFFER, B.L. (1974). Diphenamid metabolism in tomato: time course of an ozone fumigation effect. Weed Sci., 22, 205-10.   26,144

HODGSON, R.H., FREAR, D.S., SWANSON, H.R. and REGAN, L.A. (1973). Alteration of diphenamid metabolism in tomato by ozone. Weed Sci., 21, 542-9. 26

HOFFMAN, G.J., MAAS, E.V. and RAWLINS, S.L. (1973). Salinity-ozone interactive effects on yield and water relations of Pinto bean. J. Environ. Qual., 2, 148-52.   26

HOFSTRA, G. and HALL, R. (1971). Injury on roadside trees: leaf injury on pine and white cedar in relation to foliar levels of sodium and chloride. Can. J. Bot., 49, 613-22.   95

HOFSTRA, G. and LUMIS G.P. (1975). Levels of deicing salt producing injury on apple trees. Can. J. Pl. Sci., 55, 113-5.   153

HOFSTRA, G. and ORMROD, D.P. (1977). Ozone and sulphur dioxide interaction in white bean and soybean. Can. J. Pl. Sci., 57, 1193-8.   24,116

HOLMES, F.W. (1961). Salt injury to trees. Phytopathology, 51, 712-8. 97,194

HOLMES, F.W. and BAKER, J.H. (1966). Salt injury to trees. II. Sodium and chloride in roadside sugar maples in Massachusetts. Phytopathology, 56, 633-6.   97

HOOKER, W.J., YANG, T.C. and POTTER, H.S. (1973). Air pollution injury of potato in Michigan. Am. Potato J., 50, 151-61.   133

HORSMAN, D.C. and WELLBURN, A.R. (1975). Synergistic effect of SO_2 and NO_2 polluted air upon enzyme activity in pea seedlings. Environ. Pollut., 8, 123-33.   101

HORSMAN, D.C. and WELLBURN, A.R. (1976). Guide to the metabolic and biochemical effects of air pollutants on higher plants. In: Mansfield, T.A., Ed., Effects of Air Pollutants on Plants. Soc. Exptl. Biol., Sem. Ser. 1, pp. 185-99, Cambridge Univ. Press, Cambridge.   101

HORTON, B.D. and EDWARDS, J.H. (1976). Diffusive resistance rates and stomatal aperture of peach seedlings as affected by aluminum concentration. HortScience, 11, 591-3.   65

HOUSTON, D.B. and STAIRS, G.R. (1973). Genetic control of sulfur dioxide and ozone tolerance in eastern white pine. Forest Sci., 19, 267-71.
                                    159,188

HOWELL, R.K. (1970). Influence of air pollution on quantities of caffeic acid isolated from leaves of Phaseolus vulgaris. Phytopathology, 60, 1626-9.   106

HOWELL, R.K. (1974). Phenols, ozone, and their involvement in pigmentation and physiology of plant injury. In: Dugger, M., Ed., Air Pollution Effects on Plant Growth, ACS Symp. Ser. 3, pp. 94-105, Amer. Chem. Soc., Washington, D.C.   5,106

HOWELL, R.K. and KREMER, D.F. (1973). The chemistry and physiology of pigmentation in leaves injured by air pollution. J. Environ. Qual., 2, 434-8.   106,115

HUANG, G.-Y., BAZZAZ, F.A. and VANDERHOEF, L.N. (1974). The inhibition of
    soybean metabolism by cadmium and lead. Pl. Physiol., 54, 122-4.   108

HULL, H.M. and WENT, F.W. (1952). Life processes of plants as affected by
    air pollution. Proc. 2nd. Nat. Air Pollut. Symp., 122-218.   146

HULL, H.M., WENT, F.W. and YAMADA, N. (1954). Fluctuations in sensitivity
    of the Avena test due to air pollutants. Pl. Physiol., 29, 182-7.   44

HUNTER, R. and WELKIE, G.W. (1977). Growth of copper-treated corn roots
    as affected by EDTA, IAA, succinic acid-2, 2-dimethyl hydrazide,
    vitamins and potassium. Environ. Exp. Bot., 17, 19-26.   70

ISERMANN, K. (1977). A method to reduce contamination and uptake of lead
    by plants from car exhaust gases. Environ. Pollut., 12, 199-203.   72

JACOBS, L.W., KEENEY, D.R. and WALSH, L.M. (1970). Arsenic residue
    toxicity to vegetable crops grown on Plainfield sand. Agron. J., 62,
    588-91.   66

JACOBSON, J.S. and COLAVITO, L.J. (1976). The combined effect of sulfur
    dioxide and ozone on bean and tobacco plants. Environ. Exp. Bot.,
    16, 277-85.   117

JACOBSON, J.S. and HILL, A.C., Eds. (1970). Recognition of Air Pollution
    Injury to Vegetation: A Pictorial Atlas. Air Pollut. Contr. Ass.,
    Pittsburgh, Pa.   4,7,49
    123,124,127,129,130,136,138,139,155,159,160,162,165,168,176,180

JACOBSON, J.S., WEINSTEIN, L.H., MCCUNE, D.C. and HITCHCOCK, A.E. (1966).
    The accumulation of fluorine by plants. J. Air Pollut. Contr. Ass.,
    16, 412-7.   16

JAGER, H.-J. and KLEIN, H. (1977). Biochemical and physiological detec-
    tion of sulfur dioxide injury to pea plants (Pisum sativum). J. Air
    Pollut. Contr. Ass., 27, 464-6.   107,131

JANAKIRAMAN, R and HARNEY, P.M. (1976). Effects of ozone on meiotic
    chromosomes of Vicia faba. Can. J. Gen. Cyt., 18, 727-30.   109

JENSEN, K.F. (1973). Response of nine forest tree species to chronic
    ozone fumigation. Pl. Dis. Rep., 57, 914-7.   188

JENSEN, K.F. and DOCHINGER, L.S. (1974). Responses of hybrid poplar
    cuttings to chronic and acute levels of ozone. Environ. Pollut., 6,
    289-95.   188

JENSEN, K.F. and MASTERS, R.G. (1975). Growth of six woody species fumiga-
    ted with ozone. Pl. Dis. Rep., 59, 760-2.   188

JOHN, M.K. (1972a). Mercury uptake from soil by various plant species.
    Bull. Environ. Contam. Toxicol., 8, 77-80.   73,147

JOHN, M.K. (1972b). Uptake of soil-applied cadmium and its distribution
    in radishes. Can. J. Pl. Sci., 52, 715-9.   69,137

JOHN, M.K. (1973). Cadmium uptake by eight food crops as influenced by
    various soil levels of cadmium. Environ. Pollut., 4, 7-15.   69,147

JOHN, M.K. (1976). Interrelationships between plant cadmium and uptake of
    some other elements from culture solutions by oats and lettuce.
    Environ. Pollut., 11, 85-95.   67,129

JOHN, M.K. and VAN LAERHOVEN, C. (1972). Lead uptake by lettuce and oats as affected by lime, nitrogen, and sources of lead. J. Environ. Qual., 1, 169-71. 70,72

JOHN, M.K. and VAN LAERHOVEN, C.J. (1976a). Differential effects of cadmium on lettuce varieties. Environ. Pollut., 10, 163-73. 129

JOHN, M.K. and VAN LAERHOVEN, C.J. (1976b). Effects of sewage sludge composition, application rate, and lime regime on plant availability of heavy metals. J. Environ. Qual., 5, 246-51. 77

JOHN, M.K., VAN LAERHOVEN, C.J. and CHUAH, H.H. (1972). Factors affecting plant uptake and phytotoxicity of cadmium added to soils. Environ. Sci. Technol., 6, 1005-9. 68,129

JOHNELS, A.S. and WESTERMARK, T. (1969). Mercury contamination of the environment in Sweden. In: Miller, M.W. and Berg, G.C., Eds., Chemical Fallout: Current Research on Persistent Pesticides, pp. 221-41, Charles C. Thomas, Chicago, Ill. 84

JOHNSON, H.M. and APPLEGATE, H.G. (1962). Growth of fluoride treated Kalanchoe pinnata plants. Phyton, 18, 57-8. 49

JOHNSON, F., ALLMENDINGER, D.F., MILLER, V.L. and GOULD, C.J. (1950). Leaf scorch of gladiolus caused by atmospheric fluoric effluents. Phytopathology, 40, 239-46. 17,174

JONAS, H. (1969). Action of air pollutants on the biosynthesis of secondary plant products. Econ. Bot., 23, 210-4. 99

JONES, J.L. (1963). Ozone damage: protection for plants. Science, 140, 1317-8. 46

JUHREN, M., NOBLE, W. and WENT, F.W. (1957). The standardization of Poa annua as an indicator of smog concentrations. I. Effects of temperature, photoperiod, and light intensity during growth of test plants. Pl. Physiol., 32, 576-86. 34,35,184

KATS, G., THOMPSON, C.R. and KUBY, W.C. (1976). Improved ventilation of open top chambers. J. Air Pollut. Contr. Ass., 26, 1089-90. 63

KAUDY, J.C., BINGHAM, F.T., MCCOLLOCH, R.C., LIEBIG, G.F. and VANSELOW, A.P. (1955). Contamination of citrus foliage by fluorine from air pollution in major California citrus areas. Proc. Am. Soc. Hort. Sci., 65, 121-7. 158

KEARNEY, P.C., NASH, R.G. and ISENSEE, A.R. (1969). Persistence of pesticide residues in soils. In: Miller, M.W. and Berg, G.C., Eds., Chemical Fallout: Current Research on Persistent Pesticides, pp. 54-67, Charles C. Thomas, Chicago, Ill. 85

KENDER, W.J. and CARPENTER, S.G. (1974). Susceptibility of grape cultivars and selections to oxidant injury. Fruit Var. J., 28, 59-61. 160

KENDER, W.J. and SHAULIS, N.J. (1976). Vineyard management practices influencing oxidant injury in 'Concord' grapevines. J. Am. Soc. Hort. Sci., 101, 129-32. 39,160,161

KENDER, W.J. and SPIERINGS, F.H.F.G. (1975). Effects of sulfur dioxide, ozone, and their interactions on 'Golden Delicious' apple trees. Neth. J. Pl. Pathol., 81, 149-51. 24,151

KENDER, W.J., TASCHENBERG, E.F. and SHAULIS, N.J. (1973). Benomyl protection of grapevines from air pollution injury. HortScience, 8, 396-8.
44,160

KENDRICK, J.B., JR., DARLEY, E.F. and MIDDLETON, J.T. (1962). Chemotherapy for oxidant and ozone induced plant damage. Int. J. Air Wat. Pollut., 6, 391-402.
43,119

KENDRICK, J.B., JR., MIDDLETON, J.T. and DARLEY, E.F. (1953). Predisposing effects of air temperature and nitrogen supply upon injury to some herbaceous plants fumigated with peroxides derived from olefins. Phytopathology, 43, 588 (Abs.).
35,139

KENDRICK, J.B., JR., MIDDLETON, J.T. and DARLEY, E.F. (1954). Chemical protection of plants from ozonated olefin (smog) injury. Phytopathology, 44, 494 (Abs.).
42,119

KHATAMIAN, H., ADEDIPE, N.O. and ORMROD, D.P. (1973). Soil-plant-water aspects of ozone phytotoxicity in tomato plants. Pl. Soil, 38, 531-41.
37,143

KIRKHAM, M.B. (1977a). Growth of tulips treated with sludge containing dewatering chemicals. Environ. Pollut., 13, 11-20.
77,78

KIRKHAM, M.B. (1977b). Elemental composition of sludge-fertilized chrysanthemums. J. Am. Soc. Hort. Sci., 102, 352-4.
171

KIRKHAM, M.B. and KEENEY, D.R. (1974). Air pollution injury of potato plants grown in a growth chamber. Pl. Dis. Rep., 58, 304-6.
134

KIRKPATRICK, H.C., THOMPSON, J.M. and EDWARDS, J.H. (1975). Effects of aluminum concentration on growth and chemical composition of peach seedlings. HortScience, 10, 132-4.
65

KLINGAMAN, G.L. and LINK, C.B. (1975). Reduction air pollution injury to foliage to Chrysanthemum morifolium Ramat. using tolerant cultivars and chemical protectants. J. Am. Soc. Hort. Sci., 100, 173-5.
44,46,112,169,170

KNABE, W. (1976). Effects of sulfur dioxide on terrestrial vegetation. AMBIO, 5, 213-8.
13,21

KNUDSON, L.L., TIBBITTS, T.W. and EDWARDS, G.E. (1977). Measurement of ozone injury by determination of leaf chlorophyll concentration. Pl. Physiol., 60, 606-8.
49

KOEPPE, D.E. and MILLER, R.J. (1970). Lead effects on corn mitochondrial respiration. Science, 167, 1376-7.
108

KOIWAI, A. and KISAKI, T. (1973). Mixed function oxidase inhibitors protect plants from ozone injury. Agr. Biol. Chem., 37, 2449-50.
46

KORITZ, H.G. and WENT, F.W. (1953). The physiological action of smog on plants. I. Initial growth and transpiration studies. Pl. Physiol., 28, 50-62.
33,139,142

KOUKOL, J. and DUGGER, W.M., JR. (1967). Anthocyanin formation as a response to ozone and smog treatment in Rumex crispus L. Pl. Physiol., 42, 1023-4.
106

KRAAL, H. and ERNST, W. (1976). Influence of copper high tension lines on plants and soils. Environ. Pollut., 11, 131-5.
70

KRATKY, B.A., FUKUNAGA, E.T., HYLIN, J.W. and NAKANO, R.T. (1974).
Volcanic air pollution: deleterious effects on tomatoes. J. Environ.
Qual., 3, 138-40.   146

KRAUSE, G.H.M. and KAISER, H. (1977). Plant response to heavy metals and
sulphur dioxide. Environ. Pollut., 12, 63-71.   26

LACASSE, N.L. and RICH, A.E. (1964). Maple decline in New Hampshire.
Phytopathology, 54, 1071-5.   97,194

LAGERWERFF, J.V. (1971). Uptake of cadmium, lead and zinc by radish from
soil and air. Soil Sci., 111, 129-33.   75,138

LAGERWERFF, J.V. and SPECHT, A.W. (1970). Contamination of roadside soil
and vegetation with cadmium, nickel, lead, and zinc. Environ. Sci.
Technol., 4, 583-6.   74

LAHANN, R.W. (1976). Molybdenum hazard in land disposal of sewage sludge.
Wat., Air, Soil Pollut., 6, 3-8.   74

LANGE, A.H. and CRANE, J.C. (1967). The phytotoxicity of several herbicides to deciduous fruit tree seedlings. Proc. Am. Soc. Hort. Sci.,
90, 47-55.   85

LANGILLE, A.R. (1976). One season's salt accumulation in soil and trees
adjacent to a highway. HortScience, 11, 575-6.   95

LANPHEAR, F.O. and SOULE, O.H. (1970). Injury to city plants from industrial emissions of herbicides. HortScience, 5, 215-7.   86,87,88,89

LEBLANC, F., RAO, D.N. and COMEAU, G. (1972). The epiphytic vegetation
of Populus balsamifera and its significance as an air pollution indicator in Sudbury, Ontario. Can. J. Bot., 50, 519-28.   52

LEDBETTER, M.C., ZIMMERMAN, P.W. and HITCHCOCK, A.E. (1959). The histological effects of ozone on plant foliage. Contrib. Boyce Thompson
Inst., 20, 275-82.   109

LEE, R.E., Ed. (1976). Air Pollution from Pesticides and Agricultural
Processes. CRC Press, Inc., Cleveland, Ohio.   81

LEFFLER, H.R. and CHERRY, J.H. (1974). Destruction of enzymatic activities
of corn and soybean leaves exposed to ozone. Can. J. Bot., 52, 1233-8. 101

LEISTRA, M. and DEKKERS, W.A. (1976). Computed leaching of pesticides
from soil under field conditions. Wat., Air, Soil Pollut., 5, 491-
500.   82,136

LEONE, I.A. (1976). Response of potassium-deficient tomato plants to
atmospheric ozone. Phytopathology, 66, 734-6.   40,143

LEONE, I.A. and BRENNAN, E. (1969). Sensitivity of begonias to air pollution. Hort. Res., 9, 112-6.   37,164

LEONE, I.A. and BRENNAN, E. (1970). Ozone toxicity in tomato as modified
by phosphorus nutrition. Phytopathology, 60, 1521-4.   40,143

LEONE, I.A. and BRENNAN E. (1972a). Modification of sulfur dioxide injury
to tobacco and tomato by varying nitrogen and sulfur nutrition. J.
Air Pollut. Contr. Ass., 22, 544-7.   40,48,145

220

LEONE, I.A. and BRENNAN, E. (1972b). Sulfur nutrition as it contributes to the susceptibility of tobacco and tomato to SO_2 injury. Atmos. Environ., 6, 259-66.   40,145

LEONE, I.A. and GREEN, D. (1974). A field evaluation of air pollution effects on petunia and potato cultivars in New Jersey. Pl. Dis. Rep., 58, 683-7.   133

LEONE, I.A., BRENNAN, E. and DAINES, R.H. (1956). Atmospheric fluoride: its uptake and distribution in tomato and corn plants. Pl. Physiol., 31, 329-33.   17,126,145

LERMAN, S.L. and DARLEY, E.F. (1975). Particulates. In: Mudd, J.B. and Kozlowski, T.T., Eds., Responses of Plants to Air Pollution, pp. 141-58, Academic Press, Inc., New York.   20

LESLEY, J.W. and TAYLOR, O.C. (1973). Temperature and air pollution effects on early fruit production of F_2 tomato hybrids. Calif. Agr., 1973-2, 13-14.   141

LIBERA, W., ZIEGLER, H. and ZIEGLER, I. (1973). Förderung der Hill-Reaktion und der CO_2-Fixierung in isolierten Spinatchloroplasten durch niedere Sulfitkönzentrationen. Planta, 109, 269-79.   104

LIBERA, W., ZIEGLER, I. and ZIEGLER, H. (1975). The action of sulfite on the HCO_3-fixation and the fixation pattern of isolated chloroplasts and leaf tissue slices. Z. Pflanzenphysiol., 74, 420-33.   104

LIKENS, G.E., BORMANN, F.H. and JOHNSON, N.M. (1972). Acid rain. Environment, 14, 33-40.   21

LOCKYER, D.R., COWLING, D.W. and JONES, J.H.P. (1976). A system for exposing plants to atmospheres containing low concentrations of sulphur dioxide. J. Exp. Bot., 27, 397-409.   55

LOVE, J.L. and DONELLY, R.G.C. (1976). Residues of 2,4-dichlorophenoxy-acetic acid (2,4-D) on apricots. N.Z. J. Exp. Agr., 4, 369-71.   86

LUMIS, G.P., HOFSTRA, G. and HALL, R. (1973). Sensitivity of roadside trees and shrubs to aerial drift of deicing salt. HortScience, 8, 475-7.   95,96,195

LUMIS, G.P., HOFSTRA, G. and HALL, R. (1976). Roadside woody plant susceptibility to sodium and chloride accumulation during winter and spring. Can. J. Pl. Sci., 56, 853-9.   95

LUNIN, J. (1971). Agricultural wastes and environmental pollution. Adv. Environ. Sci. Technol., 2, 215-61.   81

LUNIN, J. and STEWART, F.B. (1961). The effects of soil salinity on azaleas and camellias. Proc. Am. Soc. Hort. Sci., 77, 528-32.   194

LUNT, O.R., KOHL, H.C., JR. and KOFRANEK, A.M. (1956). The effect of bicarbonate and other constituents of irrigation water on the growth of azaleas. Proc. Amer. Soc. Hort. Sci., 68, 537-44.   194

LUNT, O.R., KOHL, H.C., JR. and KOFRANEK, A.M. (1957). Tolerance of azaleas and gardenias to salinity conditions and boron. Proc. Am. Soc. Hort. Sci., 69, 543-48.   193

LUTTGE, U., OSMOND, C.B., BALL, E., BRINCKMANN, E. and KINZE, G. (1972).
Bisulfite compounds as metabolic inhibitors: nonspecific effects on
membranes. Pl. Cell Physiol., 13, 505-14.   100

MACLEAN, A.J. (1976). Cadmium in different plant species and its availability in soils as influenced by organic matter and additions of
lime, P, Cd and Zn. Can. J. Soil Sci., 56, 129-38.   68

MACLEAN, D.C. (1973). Fluoride phytotoxicity as affected by relative
humidity. In: Proc. 3rd Int. Clean Air Congr., pp. A143-5, Int.
Union Air Pollut. Prev. Assocs. VDI-Verlag, Dusseldorf.   38,174,175

MACLEAN, D.C. and SCHNEIDER, R.E. (1976). Photochemical oxidants in Yonkers,
New York: effects on yield of bean and tomato. J. Environ. Qual., 5,
75-8.   116,140

MACLEAN, D.C., SCHNEIDER, R.E. and MCCUNE, D.C. (1976). Fluoride susceptibility of tomato plants as affected by magnesium nutrition. J. Am.
Soc. Hort. Sci., 101, 347-52.   41,146

MACLEAN, D.C., SCHNEIDER, R.E. and MCCUNE, D.C. (1977). Effects of chronic
exposure to gaseous fluoride on yield of field-grown bean and tomato
plants. J. Am. Soc. Hort. Sci., 102, 297-9.   17,121,145

MACLEAN, D.C., MCCUNE, D.C., WEINSTEIN, L.H., MANDL, R.H. and WOODRUFF, G.N.
(1968). Effects of acute hydrogen fluoride and nitrogen dioxide exposures on citrus and ornamental plants of central Florida. Environ.
Sci. Technol., 2, 444-9.   157,158,192,193

MACLEAN, D.C., ROARK, O.F., FOLKERTS, G. and SCHNEIDER, R.E. (1969).
Influence of mineral nutrition on the sensitivity of tomato plants
to hydrogen fluoride. Environ. Sci. Technol., 3, 1201-4.   40

MAJERNIK, O. and MANSFIELD, T.A. (1971). Effects of SO_2 pollution on stomatal movements in Vicia faba. Phytopath. Z., 71, 123-8.   30

MAJERNIK, O. and MANSFIELD, T.A. (1972). Stomatal responses to raised
atmospheric CO_2 concentrations during exposure of plants to SO_2
pollution. Environ. Pollut., 3, 1-7.   30

MALHOTRA, S.S. and HOCKING, D. (1976). Biochemical and cytological effects
of sulphur dioxide on plant metabolism. New Phytol., 76, 227-37.   13

MALONE, C., KOEPPE, D.E. and MILLER, R.J. (1974). Localization of lead
accumulated by corn plants. Pl. Physiol., 53, 388-94.   72

MANDL, R.H., WEINSTEIN, L.H. and KEVENY, M. (1975). Effects of hydrogen
fluoride and sulphur dioxide alone and in combination on several
species of plants. Environ. Pollut., 9, 133-43.   24,25

MANDL, R.H., WEINSTEIN, L.H., MCCUNE, D.C. and KEVENY, M. (1973). A cylindrical, open-top chamber for the exposure of plants to air pollutants in
the field. J. Environ. Qual., 2, 371-6.   62

MANN, S.K. (1977). Cytological and genetical effects of dithane fungicide
on Allium cepa. Environ. Exp. Bot., 17, 7-12.   90,130

MANNING, W.J. (1975). Interactions between air pollutants and fungal,
bacterial and viral plant pathogens. Environ. Pollut., 9, 87-90.   27

MANNING, W.J. and FEDER, W.A. (1976). Effects of ozone on economic plants. In: Mansfield, T.A., Ed., Effects of Air Pollutants on Plants. Soc. Exptl. Bot. Sem. Ser. 1, pp. 47-60, Cambridge Univ. Press, Cambridge.
44,120,141

MANNING, W.J. and VARDARO, P.M. (1973). Suppression of oxidant injury on beans by systemic fungicides. Phytopathology, 63, 1415-6. 120

MANNING, W.J., FEDER, W.A. and PERKINS, I. (1970). Ozone injury increases infection of geranium leaves by Botrytis cinerea. Phytopathology, 60, 669-70. 27,173

MANNING, W.J., FEDER, W.A. and PERKINS, I. (1972a). Effects of Botrytis and ozone on bracts and flowers of poinsettia cultivars. Pl. Dis. Rep., 56, 814-6. 28

MANNING, W.J., FEDER, W.A. and PERKINS, I. (1972b). Sensitivity of spinach cultivars to ozone. Pl. Dis. Rep., 56, 832-3. 138

MANNING, W.J., FEDER, W.A. and PERKINS, I. (1973a). Response of poinsettia cultivars to several concentrations of ozone. Pl. Dis. Rep., 57, 774-5. 180

MANNING, W.J., FEDER, W.A. and VARDARO, P.M. (1973b). Reduction of chronic injury on poinsettia by benomyl. Can. J. Pl. Sci., 53, 833-5. 44,181

MANNING, W.J., FEDER, W.A. and VARDARO, P.M. (1973c). Benomyl in soil and responses of Pinto bean plants to repeated exposures to a low level of ozone. Phytopathology, 63, 1539-40. 119

MANNING, W.J., FEDER, W.A. and VARDARO, P.M. (1974). Suppression of oxidant injury by benomyl: effects on yields of bean cultivars in the field. J. Environ. Qual, 3, 1-3. 53,119

MANNING, W.J., FEDER, W.A., PAPIA, P.M. and PERKINS, I. (1971a). Effect of low levels of ozone on growth and susceptibility of cabbage plants to Fusarium oxysporum f. sp. conglutinans. Pl. Dis. Rep., 55, 47-9. 28

MANNING, W.J., FEDER, W.A., PAPIA, P.M. and PERKINS, I. (1971b). Influence of foliar ozone injury on root development and root surface fungi of pinto bean plants. Environ. Pollut., 1, 305-12. 28

MANNING, W.J., FEDER, W.A., PERKINS, I. and GLICKMAN, M. (1969). Ozone injury and infection of potato leaves by Botrytis cineria. Pl. Dis. Rep., 53, 691-3. 27,134

MANSFIELD, T.A. (1973). The role of stomata in determining the response of plants to air pollutants. Comments. Pl. Sci., 2, 11-20. 29

MANSFIELD, T.A., Ed. (1976). Effects of Air Pollutants on Plants. Soc. Exptl. Biol. Sem. Ser. 1, Cambridge Univ. Press, Cambridge. 4

MANSFIELD, T.A. and MAJERNIK, O. (1970). Can stomata play a part in protecting plants against air pollution? Environ. Pollut., 1, 149-54. 30

MARKOWSKI, A. and GRZESIAK, S. (1974). Influence of sulphur dioxide and ozone on vegetation of bean and barley plants under different soil moisture conditions. Bull. Acad. Polonaise Sci., Ser. Sci. Biol., 22, 875-87. 117

MARKOWSKI, A., GRZESIAK, S. and SCHRAMEL, M. (1974). Susceptibility of six species of cultivated plants to sulphur dioxide under optimum soil moisture and drought conditions. Bull. Acad. Polonaise Sci., Ser. Sci. Biol., 22, 889-98. 38,120

MARTIN, A. and BARBER, F.R. (1971). Some measurements of loss of atmospheric sulphur dioxide near foliage. Atmos. Environ., 5, 345-52. 31

MASARU, N., SYOZO, F and SABURO, K. (1976). Effects of exposure to various injurious gases on germination of lily pollen. Environ. Pollut., 11, 181-7. 52,175,176

MATSUMOTO, H., WAKIUCHI, N. and TAKAHASHI, E. (1968). Changes of sugar levels in cucumber leaves during ammonium toxicity. Physiol. Pl. 21, 1210-6. 78

MATSUSHIMA, J. and BREWER, R.F. (1972). Influence of sulfur dioxide and hydrogen fluoride as a mix or reciprocal exposure on citrus growth and development. J. Air Pollut. Contr. Ass., 22, 710-3. 24,60,159

MAVRODINEANU, R., GWIRTSMAN, J., MCCUNE, D.C. and PORTER, C.A. (1962). Summary of procedures used in the controlled fumigation of plants with volatile fluorides and in the determination of fluorides in air, water, and plant tissues. Contrib. Boyce Thompson Inst., 21, 453-64. 63

MAXIE, E.C., FARNHAM, D.S., MITCHELL, F.G., SOMMER, N.F., PARSONS, R.A., SNYDER, R.G. and RAE, H.L. (1973). Temperature and ethylene effects on cut flowers of carnations (Dianthus caryophyllus L.). J. Am. Soc. Hort. Sci., 98, 568-72. 166

MAYAK, S. and KOFRANEK, A.M. (1976). Altering the sensitivity of carnation flowers (Dianthus caryophyllus L.) to ethylene. J. Am. Soc. Hort. Sci., 101, 503-6. 35,38,45,167,168

MAYAK, S., VAADIA, Y. and DILLEY, D.R. (1977). Regulation of senescence in carnation (Dianthus caryophyllus) by ethylene. Mode of action. Pl. Physiol., 59, 591-3. 167

MAYNARD, D., BARKER, A. and LACHMANN, W. (1966). Ammonium-induced stem and leaf lesions of tomato plants. Proc. Am. Soc. Hort. Sci., 88, 516-20. 78,79

MCCALLAN, S.E.A., HARTZELL, A. and WILCOXON, F. (1936). Hydrogen sulphide injury to plants. Contrib. Boyce Thompson Inst., 8, 189-97. 19,81

MCCUNE, D. (1969). Fluoride criteria for vegetation reflect the diversity of the plant kingdom. Environ. Sci. Technol., 3, 720,7,8,31,2,6. 18

MCCUNE, D.C. (1973). Summary and synthesis of plant toxicology. In: Naegele, J.A., Ed., Air Pollution Damage to Vegetation, Adv. Chem. Ser. 122, pp. 48-62, Amer. Chem. Soc., Washington, D.C. 7

MCCUNE, D.C. and WEINSTEIN, L.H. (1971). Metabolic effects of atmospheric fluorides on plants. Environ. Pollut., 1, 169-74. 107

MCCUNE, D.C., HITCHCOCK, A.E. and WEINSTEIN, L.H. (1967). Effect of mineral nutrition on the growth and sensitivity of gladiolus to hydrogen fluoride. Contrib. Boyce Thompson Inst., 23, 295-9. 40,41,174

MCCUNE, D.C., MACLEAN, D.C. and SCHNEIDER, R.E. (1976). Experimental
approaches to the effects of airborne fluoride on plants. In:
Mansfield, T.A., Ed., Effects of Air Pollutants on Plants. Soc.
Exptl. Biol., Sem. Ser. I, pp. 31-46, Cambridge Univ. Press, Cambridge. 61

MCCUNE, D.C., SILBERMAN, D.H., MANDL, R.H., WEINSTEIN, L.H., FREUDENTHAL,
P.C. and GIARDIA, P.A. (1977). Studies on the effects of saline
aerosols of cooling tower origin on plants. J. Air Pollut. Contr.
Ass., 27, 319-24.   94,123,194

MCILVEEN, W.D., SPOTTS, R.A. and DAVIS, D.D. (1975). The influence of
soil zinc on nodulation, mycorrhizae, and ozone-sensitivity of Pinto
bean. Phytopathology, 65, 645-7.   26

MCKEEN, C.D., FULTON, J.M. and FINDLAY, W.I. (1973). Fleck and acidosis
of potatoes in southwestern Ontario. Can. Pl. Dis. Survey, 53, 150-2. 134

MCNULTY, I.B. and NEWMAN, D.W. (1957). Effects of atmospheric fluoride on
the respiration rate of bush bean and gladiolus leaves. Pl. Physiol.,
32, 121-4.   107,122,174

MCNULTY, I.B. and NEWMAN, D.W. (1961). Mechanism(s) of fluoride induced
chlorosis. Pl. Physiol., 36, 385-8.   108

MENSER, H.A. and HEGGESTAD, H.E. (1964). A facility for ozone fumigation
of plant materials. Crop Sci., 4, 103-5.   55

METCALF, R.L., REYNOLDS, H.T., WINTON, M and FUKTO, T.R. (1959). Effects
of temperature and plant species upon the rates of metabolism of
systemically applied Di-syston. J. Econ. Ent., 52, 435-9.   90

MIDDLETON, J.T. (1956). Response of plants to air pollution. J. Air
Pollut. Contr. Ass., 6, 1-4.   36,38,114,128,139

MIDDLETON, J.T. (1965). The presence, persistence and removal of pesticides
in air. In: Chichester, C.O., Ed., Research in Pesticides, pp. 191-7,
Academic Press, New York.   82

MIDDLETON, J.T., DARLEY, E.F. and BREWER, R.F. (1958). Damage to vegetation from polluted atmospheres. J. Air Pollut. Contr. Ass., 8, 9-15. 4,142

MIDDLETON, J.T., KENDRICK, J.B., JR. and SCHWALM, H.W. (1950). Injury to
herbaceous plants by smog or air pollution. Pl. Dis. Rep., 34, 245-52.
                                                                11,49,116

MILES, J.R.W. (1968). Arsenic residues in agricultural soils of southwestern Ontario. J. Agr. Food Chem., 16, 120-2.   66

MILLER, J.E. and MILLER, G.W. (1974). Effects of fluoride on mitochondrial
activity of higher plants. Physiol. Pl., 32, 115-21.   107

MILLER, J.E., HASSETT, J.J. and KOEPPE, D.E. (1977). Interactions of lead
and cadmium on metal uptake and growth of corn plants. J. Environ.
Qual., 6, 18-20.   75

MILLER, P.M. and RICH, S. (1967). Soot damage to greenhouse plants. Pl.
Dis. Rep., 51, 712.   21

MILLER, P.M. and RICH, S. (1968). Ozone damage on apples. Pl. Dis. Rep.,
52, 730-1.   30,150

MILLER, P.M., TOMLINSON, H. and TAYLOR, G.S. (1976). Reducing severity of ozone damage to tobacco and beans by combining benomyl or carboxin with contact nematicides. Pl. Dis. Rep., 60, 433-6. 26,120

MILLER, V.L., JOHNSON, F. and ALLMENDINGER, D.R. (1948). Fluorine analysis of Italian prune foliage affected by marginal scorch. Phytopathology, 38,30-7. 17

MINCHIN, F.R. and PATE, J.S. (1975). Effects of water, aeration, and salt regime on nitrogen fixation in a nodulated legume-definition of an optimum root environment. J. Exp. Bot., 26, 60-9. 108

MITCHELL, C.D. and FRETZ, T.A. (1977). Cadmium and zinc toxicity in white pine, red maple, and Norway spruce. J. Am. Soc. Hort. Sci., 102,81-4. 193

MOHAMED, A. (1968). Cytogenetic effects of hydrogen fluoride treatment in tomato plants. J. Air Pollut. Contr. Ass., 18, 395-8. 111

MOHAMED, A.H. (1970). Chromosomal changes in maize induced by hydrogen fluoride gas. Can. J. Gen. Cyt., 12, 614-20. 111

MONK, R. and PETERSON, H.B. (1962). Tolerance of some trees and shrubs to saline conditions. Proc. Amer. Soc. Hort. Sci., 81, 556-61. 193

MOTTO, H.L., DAINES, R.H., CHILKO, D.M. and MOTTO, C.K. (1970). Lead in soils and plants: Its relationship to traffic volume and proximity to highways. Environ. Sci. Technol., 4, 231-7. 71

MOYER, J., COLE, H., JR. and LACASSE, N.L. (1974a). Reduction of ozone injury on Poa annua by benomyl and thiophanate. Pl. Dis. Rep., 58, 41-4. 44,185

MOYER, J.W., COLE, H., JR. and LACASSE, N.L. (1974b). Suppression of naturally occurring oxidant injury on azalea plants by drench or foliar spray treatment with benzimidazole or oxathiin compounds. Pl. Dis. Rep., 58, 136-8. 44,189

MUDD, J.B. (1963). Enzyme inactivation by peroxyacetyl nitrate. Arch. Biochem. Biophys., 102, 59-65. 101

MUDD, J.B. (1972). Biochemical effects of some air pollutants on plants. In: Naegele, J.A., Ed., Air Pollution Damage to Vegetation, Adv. Chem. Ser. 122, pp. 31-47, Am. Chem. Soc., Washington, D.C. 99

MUDD, J.B. (1975a). Peroxyacyl nitrates. In: Mudd, J.B. and Kozlowski, T.T., Eds., Responses of Plants to Air Pollution, pp. 97-119, Academic Press, Inc., New York. 12

MUDD, J.B. (1975b). Sulfur dioxide. In: Mudd, J.B. and Kozlowski, T.T., Eds., Responses of Plants to Air Pollution, pp. 9-22, Academic Press, Inc., New York. 15

MUDD, J.B. and KOZLOWSKI, T.T., Eds. (1975). Responses of Plants to Air Pollution, Academic Press, Inc., New York.

MUKAMMEL, E.I. (1976). Review of present knowledge of plant injury by air pollution. World Met. Org. Tech. Note 147. 27 pp. 5

MUKERJI, S.K. and YANG, S.F. (1974). Phosphoenolpyruvate carboxylase from spinach leaf tissue. Inhibition by sulfite ion. Pl. Physiol., 53, 829-34. 101

226

MUMFORD, R.A., LIPKE, H., LAUFER, D.A. and FEDER, W.A. (1972). Ozone-induced changes in corn pollen. Environ. Sci. Technol., 6, 427-30.   105

MURRAY, J.J., HOWELL, R.K. and WILTON, A.C. (1975). Differential response of seventeen Poa pratensis cultivars to ozone and sulfur dioxide. Pl. Dis. Rep., 59, 852-4.   185

NATIONAL ACADEMY OF SCIENCES (1968). Effects of pesticides on fruit and vegetable physiology. Vol. 6 of Principles of Plant and Animal Pest Control. 90 pp., Washington, D.C.   80

NATIONAL ACADEMY OF SCIENCES (1977). Medical and biologic effects of environmental pollutants. Ozone and other photochemical oxidants. Committee on Medical and Biologic Effects of Environmental Pollutants. 719 pp. Washington, D.C.   10

NEIL, L.J., ORMROD, D.P. and HOFSTRA, G. (1973). Ozone stimulation of tomato stem elongation. HortScience, 8, 488-9.   48,140

NICHOLS, R. (1968). The response of carnations (Dianthus caryophyllus) to ethylene. J. Hort. Sci., 43, 335-49.   166

NOBEL, P.S. (1974). Ozone effects on chlorophylls a and b. Naturwissenschaften, 61, 80-1.   102

NOBEL, P.S. and WANG, C.-T. (1973). Ozone increases the permeability of isolated pea chloroplasts. Arch. Biochem. Biophys., 157, 388-94.   102

ODOI, N. (1976). 1976 index of plant sensitivity to pollution. Grounds Maintenance, 11, 68, 70, 72, 74-6.   6

OERTLI, J.T. (1959). Effects of salinity on susceptibility of sunflower plants to smog. Soil Sci., 87, 249-51.   38

OGATA, G. and MAAS, E.V. (1973). Interactive effects of salinity and ozone on growth and yield of garden beet. J. Environ. Qual., 2, 518-20.   26,123

OLIVA, M. and STEUBING, L. (1976). Untersuchungen über die beeinflussung von photosynthese, respiration und wasserhaushalt durch H_2S bei Spinacia oleracea. Angew. Bot., 50, 1-17.   20,64,104,139

ORDIN, L. and PROPST, B. (1962). Effect of air-borne oxidants on biological activity of indoleacetic acid. Bot. Gaz., 122, 170-5.   45

ORMROD, D.P. (1976). Sensitivity of pea cultivars to ozone. Pl. Dis. Rep., 60, 423-6.   130,131

ORMROD, D.P. (1977). Cadmium and nickel effects on growth and ozone sensitivity of pea. Wat., Air, Soil Pollut., 8, 263-70.   26,131,132

ORMROD, D.P. and ADEDIPE, N.O. (1974). Protecting horticultural plants from atmospheric pollutants: A review. HortScience, 9, 108-11.   41

ORMROD, D.P. and ADEDIPE, N.O. (1975). Experimental exposures and crop monitors to confirm air pollution. HortScience, 10, 493-4.   50

ORMROD, D.P., ADEDIPE, N.O. and BALLANTYNE, D.J. (1976). Air pollution injury to horticultural plants: A review. Hort. Abs., 46, 241-8.   6

ORMROD, D.P., ADEDIPE, N.O. and HOFSTRA, G. (1971). Responses of cucumber, onion, and potato cultivars to ozone. Can. J. Pl. Sci., 51, 283-8.   126

ORMROD, D.P., ADEDIPE, N.O. and HOFSTRA, G. (1973). Ozone effects on growth of radish plants as influenced by nitrogen and phosphorus nutrition and by temperature. Pl. Soil, 39, 437-9.   35,39,137

OSHIMA, R.J. (1973). Effect of ozone on a commercial sweet corn variety. Pl. Dis. Rep., 57, 719-23.   125

OSHIMA, R.J. (1974). A viable system of biological indicators for monitoring air pollutants. J. Air Pollut. Contr. Ass., 24, 576-8.   52

OSHIMA, R.J., TAYLOR, O.C., BRAEGELMANN, P.K. and BALDWIN, D.W. (1975). Effect of ozone on the yield and plant biomass of a commercial variety of tomato. J. Environ. Qual., 4, 463-4.   141,142

OSHIMA, R.J., BRAEGELMANN, P.K., BALDWIN, D.W., VAN WAY, V. and TAYLOR, O.C. (1977a). Responses of five cultivars of fresh market tomato to ozone: a contrast of cultivar screening with foliar injury and yield. J. Am. Soc. Hort. Sci., 102, 286-9.   112,141

OSHIMA, R.J., BRAEGELMANN, P.K., BALDWIN, D.W., VAN WAY, V. and TAYLOR, O.C. (1977b). Reduction of tomato fruit size and yield by ozone. J. Am. Soc. Hort. Sci., 102, 289-93.   142

OTTO, H.W. and DAINES, R.H. (1969). Plant injury by air pollutants: influence of humidity on stomatal apertures and plant response to ozone. Science, 163, 1209-10.   37

PACK, M.R. (1966). Response of tomato fruiting to hydrogen fluoride as influenced by calcium nutrition. J. Air Pollut. Contr. Ass., 16, 541-4.   41,145

PACK, M.R. (1971a). Effects of hydrogen fluoride on bean reproduction. J. Air Pollut. Contr. Ass., 21, 133-7.   121,122

PACK, M.R. (1971b). Effects of hydrogen fluoride on production and organic reserves of bean seed. Environ. Sci. Technol., 5, 1128-32.   121,122

PACK, M.R. (1972). Response of strawberry fruiting to hydrogen fluoride fumigation. J. Air Pollut. Contr. Ass., 22, 714-7.   162,163

PACK, M.R. and SULZBACH, C.W. (1976). Response of plant fruiting to hydrogen fluoride fumigation. Atmos. Environ. 10, 73-81.   147

PACK, M. and WILSON, A. (1967). Influence of hydrogen fluoride fumigation on acid soluble phosphorus compounds in bean seedlings. Environ. Sci. Technol., 1, 1011-3.   108

PAGE, A.L. and GANJE, T.J. (1970). Accumulations of lead in soils for regions of high and low motor vehicle traffic density. Environ. Sci. Technol., 4, 140-2.   71

PAGE, A.L., BINGHAM, F.T. and NELSON, C. (1972). Cadmium absorption and growth of various plant species as influenced by solution cadmium concentration. J. Environ. Qual., 1, 288-91.   67,147,148,149

PAHLICH, E. (1973). Uber den Hemmechanismus mitochondrialer Glutamat-Oxalacetat-Transaminase in SO_2-begasten Erbsen. Planta, 110, 267-8.   101

PAHLICH, E. (1975). Effect of SO_2-pollution on cellular regulation. A general concept of the mode of action of gaseous air contamination. Atmos. Environ., 9, 261-3. 101,107

PAPPLE, D.J. and ORMROD, D.P. (1977). Comparative efficacy of ozone-injury suppression by benomyl and carboxin on turfgrasses. J. Am. Soc. Hort. Sci., 102, 792-6. 185

PARRIS, G.K. (1968). Automobile exhaust fumes cause dieback of redbud. Pl. Dis. Rep., 52, 744. 20

PATEL, P.M., WALLACE, A. and MUELLER, R.T. (1976). Some effects of copper, cobalt, cadmium, zinc, nickel, and chromium on growth and mineral element concentration in chrysanthemum. J. Am. Soc. Hort. Sci., 101, 553-6. 172

PEARSON, R.G., DRUMMOND, D.B., MCILVEEN, W.D. and LINZON, S.N. (1974). PAN-type injury to tomato crops in southwestern Ontario. Pl. Dis. Rep., 58, 1105-8. 144

PELL, E.J. (1974). The impact of ozone on the bioenergetics of plant systems. In: Dugger, M., Ed., Air Pollution Effects on Plant Growth. ACS Symp. Ser. 3, pp. 106-14, Am. Chem. Soc., Washington, D.C. 101

PELL, E.J. (1976). Influence of benomyl soil treatment on pinto bean plants exposed to peroxyacetyl nitrate and ozone. Phytopathology, 66, 731-3. 44,120

PELL, E.J. and BRENNAN, E. (1973). Changes in respiration, photosynthesis, adenosine-5'-triphosphate, and total adenylate content of ozonated Pinto bean foliage as they relate to symptom expression. Pl. Physiol., 51, 378-81. 115

PELL, E.J. and BRENNAN, E. (1975). Economic impact of air pollution on vegetation in New Jersey and interpretation of its annual variability. Environ. Pollut., 8, 23-33. 7

PELLISSIER, M., LACASSE, N.L. and COLE, H., JR. (1972a). Effectiveness of benzimidazole, benomyl, and thiabendazole in reducing ozone injury to Pinto beans. Phytopathology, 62, 580-2. 44,119

PELLISSIER, M., LACASSE, N.L. and COLE, H., JR. (1972b). Effectiveness of benomyl-Folicote treatments in reducing ozone injury to Pinto beans. J. Air Pollut. Contr. Ass., 22, 722-5. 119

PELLISSIER, M., LACASSE, N.L., ERCEGOVICH, C.D. and COLE, H., JR. (1972c). Effects of hydrocarbon wax emulsion sprays in reducing visible ozone injury to Phaseolus vulgaris - 'Pinto III'. Pl. Dis. Rep., 56, 6-9. 46,120

PERCHOROWICZ, J.T. and TING, I.P. (1974). Ozone effects on plant cell permeability. Am. J. Bot., 61, 787-93. 100

PETERS, R.A. and SHORTHOUSE, M. (1967). Observations on the metabolism of fluoride in Acacia georginae and some other plants. Nature, 216, 80-1. 17

PIERSOL, J.R. and HANAN, J.J. (1975). Effect of ethylene on carnation growth. J. Am. Soc. Hort. Sci., 100, 679-81. 63,167

PILET, P.-E. (1969). Effet du fluor sur la teneur en RNA et l'activite RNA-asique des racines du Lens culinaris. C. r. hebd. Séanc. Acad. Sci., Paris, Sér. D., 269, 954-7. 108

PILET, P.-E. (1970). RNA metabolism and fluoride action. Fluoride, 3, 153-9. 108

PIPPEN, E.L., POTTER, A.L., RANDALL, V.G., NG, K.C., REUTER, F.W., III, MORGAN, A.I., JR. and OSHIMA, R.J. (1975). Effect of ozone fumigation on crop composition. J. Food Sci., 40, 672-6. 147,148

POOLE, R.T. and CONOVER, C.A. (1974). Foliar necrosis of Dracaena deremensis Engler cv. Warneckii cuttings induced by fluoride. HortScience, 9, 378-9. 172

POOVAIAH, B.W. and WIEBE, H.H. (1969). Tylosis formation in response to fluoride fumigation of leaves. Phytopathology, 59, 518-9. 111

POOVAIAH, B.W. and WIEBE, H.H. (1971). Effects of gaseous hydrogen fluoride on oxidative enzymes of Pelargonium zonale leaves. Phytopathology, 61, 1277-9. 102

POSTHUMUS, A.C. (1976). The use of higher plants as indicators for air pollution in the Netherlands. In L. Kärenlampi (ed.) Proc. Kuopio Mtg. Plant Damages Caused by Air Pollut., Kuopio, Finland, 115-20. 51

POSTHUMUS, A.C. (1977). Experimentelle untersuchungen der wirkung von ozon und peroxyacetylnitrat (PAN). VDI-Berichte Nr. 270, 153-61. 24

PRASAD, K., WEIGLE, J.L. and SHERWOOD, C.H. (1970). Variation in ozone sensitivity among Phaseolus vulgaris cultivars. Pl. Dis. Rep., 54, 1026-9. 113,115

PROCTOR, J.T.A. and ORMROD, D.P. (1977). Response of celery to ozone. HortScience, 12, 321-2. 124

PRZYBYLSKI, A. (1967). Effects of gases and vapors of SO_2, SO_3 and H_2SO_4 on fruit trees and certain harmful insects. Postepy. Nauk. roln., 14, 111-8. 28

PUERNER, M.J. and SIEGEL, S.M. (1972). The effects of mercury compounds on the growth and orientation of cucumber seedlings. Physiol. Pl. 26, 310-2. 74

PURITCH, G.S. and BARKER, A.V. (1967). Structure and function of tomato leaf chloroplasts during ammonium toxicity. Pl. Physiol., 42, 1229-38. 78

PUTH, G. and LUTTGE, U. (1973). Sulfitwirkung auf die Membranpermeabilität von Pflanzzellen: SO_3 -Hemmung des n-butanol induzierten Betacyaninefflux aus dem Gewebe roter Ruben. Biochem. Physiol. Pflanzen, 164, 195-8. 100

RAJPUT, C.B.S. and ORMROD, D.P. (1976). Response of eggplant cultivars to ozone. HortScience, 11, 462-3. 100,127,128

RAJPUT, C.B.S., ORMROD, D.P. and EVANS, W.D. (1977). The resistance of strawberry to ozone and sulfur dioxide. Pl. Dis. Rep., 61, 222-5. 138,162

REINERT, R.A. (1975). Monitoring, detecting, and effects of air pollutants on horticultural crops: sensitivity of genera and species. HortScience, 10, 495-500. 6

230

REINERT, R.A., HEAGLE, A.S. and HECK, W.W. (1975). Plant responses to pollutant combinations. In: Mudd, J.B. and Kozlowski, T.T., Eds., Responses of Plants to Air Pollution, pp. 159-77, Academic Press, Inc., New York.   23

REINERT, R.A., TINGEY, D.T. and CARTER, H.B. (1972a). Sensitivity of tomato cultivars to ozone. J. Am. Soc. Hort. Sci., 97, 149-51.   33,141,143

REINERT, R.A., TINGEY, D.T. and CARTER, H.B. (1972b). Ozone induced foliar injury in lettuce and radish cultivars. J. Am. Soc. Hort. Sci., 97, 711-4.   128,136

REINERT, R.A., HEAGLE, A.S. MILLER, J.R. and GECKELER, W.R. (1970). Field Studies of air pollution injury to vegetation in Cincinatti, Ohio. Pl. Dis. Rep., 54, 8-11.   10

RHOADS, A.F. and BRENNAN, E. (1976). Response of ornamental plants to chlorine contamination in the atmosphere. Pl. Dis. Rep., 60, 409-11.   19

RHOADS, A., TROIANO, J. and BRENNAN, E. (1973). Ethylene gas as a cause of injury to Easter lilies. Pl. Dis. Rep., 57, 1023-4.   175

RICH, S. (1964). Ozone damage to plants. Annu. Rev. Phytopath., 2, 253-66. 31

RICH, S. (1975). Interactions of air pollution and agricultural practices. In: Mudd, J.B. and Kozlowski, T.T., Eds., Plant Responses to Air Pollution, pp. 335-353, Academic Press, Inc., New York.   4

RICH, S. and TAYLOR, G.S. (1960). Antiozonants to protect plants from ozone damage. Science, 132, 150-1.   42,143

RICH, S. and TOMLINSON, H. (1968). Air pollution damage to petunias and daturas in Connecticut. Pl. Dis. Rep., 52, 732-3.   20,180

RICH, S. and TOMLINSON, H. (1974). Mechanisms of ozone injury to plants. In: Dugger, M., Ed., Air Pollution Effects on Plant Growth, Am. Chem. Soc. Symp. Ser. 3, pp. 76-82, Amer. Chem. Soc., Washington, D.C.   99

RICH, S. and TURNER, N.C. (1972). Importance of moisture on stomatal behavior of plants subjected to ozone. J. Air Pollut. Contr. Ass., 22, 718-21.   30,117

RICH, S., AMES, R. and ZUKEL, J.W. (1974). 1,4-Oxathiin derivatives protect plants against ozone. Pl. Dis. Rep., 58, 163-4.   44,120,143

RICH, S., TAYLOR, G. and TOMLINSON, H. (1969). Crop damaging periods of ambient ozone in Connecticut. Pl. Dis. Rep., 53, 969-73.   9

RICH, S., WAGGONER, P.E. and TOMLINSON, H. (1970). Ozone uptake by bean leaves. Science, 169, 79-80.   30,117

RICHARDS, B.L. and TAYLOR, O.C. (1965). Significance of atmospheric ozone as a phytotoxicant. J. Air Pollut. Contr. Ass., 15, 191-3.   10

RICHARDS, B.L., MIDDLETON, J.T. and HEWITT, W.B. (1958). Air pollution with relation to agronomic crops. V. Oxidant stipple of grape. Agron. J., 50, 559-61.   160

RIES, S.K., LARSEN, R.P. and KENWORTHY, A.L. (1963). The apparent influence of simazine on nitrogen nutrition of peach and apple trees. Weeds, 11, 270-3.   90

RIPPEL, A. and JANOVICOVA, J. (1969). Der einfluss von fluorexhalaten auf die pflanzenwelt in der umgeburg eines aluminiumwerkes. Proc. 1st Eur. Congr. Infl. Air Pollut. on Plants and Animals 1968, pp. 173-8, Pudoc, Wageningen.   17

ROBBINS, W.A. and TAYLOR, W.S. (1957). Injury to canning tomatoes caused by 2,4-D. Proc. Am. Soc. Hort. Sci., 70, 373-8.   146

ROBERTS, B.R. (1974). Foliar sorption of atmospheric sulphur dioxide by woody plants. Environ. Pollut., 7, 133-40.   31

ROBERTS, B.R. (1975). The influence of sulfur dioxide concentration on growth of potted white birch and pin oak seedlings in the field. J. Am. Soc. Hort. Sci., 100, 640-2.   48

ROBERTS, B.R. and KRAUSE, C.R. (1976). Changes in ambient SO_2 by Rhododendron and Pyracantha. HortScience, 11, 111-2.   31

ROBINSON, F.E. and MCCOY, O.D. (1965). The effect of sprinkler irrigation with saline water and rates of seeding on germination and growth of lettuce. Proc. Am. Soc. Hort. Sci., 87, 318-23.   94

ROGERS, H.H., JEFFRIES, H.E., STAHEL, E.P., HECK, W.W., RIPPERTON, L.A. and WITHERSPOON, A.M. (1977). Measuring air pollutant uptake by plants: a direct kinetic technique. J. Air Pollut. Contr. Ass., 27, 1192-7. 13,56

ROLFE, G.L. and BAZZAZ, F.A. (1975). Effect of lead contamination on transpiration and photosynthesis of loblolly pine and autumn olive. Forest Sci., 21, 33-5.   72,105

ROMNEY, E.M., WALLACE, A. and ALEXANDER, G.V. (1975). Responses of bush bean and barley to tin applied to soil and to solution culture. Pl. Soil, 42, 585-9.   74,123

ROSS, C.W., WIEBE, H.H., MILLER, G.W. and HURST, R.L. (1968). Respiratory pathway, flower color, and leaf area of gladiolus as factors in the resistance to fluoride injury. Bot. Gaz., 129, 49-52.   107,174

ROSS, R.G. and LONGLEY, R.P. (1963). Effect of fungicides on McIntosh apple trees. Can. J. Pl. Sci., 43, 497-502.   154

ROSS, R.G. and STEWART, D.K.R. (1962). Movement and accumulation of mercury in apple trees and soil. Can. J. Pl. Sci., 42, 280-5.
                                                          73,83,152,153

ROSS, R.G. and STEWART, D.K.R. (1964). Mercury residues in potatoes in relation to foliar sprays of phenyl mercury chloride. Can. J. Pl. Sci., 44, 123-5.   74,135

ROSS, R.G. and STEWART, D.K.R. (1969). Cadmium residues in apple fruit and foliage following a cover spray of cadmium chloride. Can. J. Pl. Sci., 49, 49-52.   152

ROSS, R.G., CROWE, A.D. and WEBSTER, D.H. (1970). Effect of fungicides on the performance of young McIntosh and Cortland apple trees. Can. J. Pl. Sci., 50, 529-36.   154

RUFNER, R., WITHAM, F.H. and COLE, H., JR. (1975). Ultrastructure of chloroplasts of Phaseolus vulgaris leaves treated with benomyl and ozone. Phytopathology, 65, 345-9.   109

RUNECKLES, V.C. (1974). Dosage of air pollutants and damage to vegetation. Environ. Conserv., 1, 305-8.   56

RUNECKLES, V.C. and RESH, H.M. (1975a). The assessment of chronic ozone injury to leaves by reflectance spectrophotometry. Atmos. Environ., 9, 447-52.   49

RUNECKLES, V.C. and RESH, H.M. (1975b). Effects of cytokinins on responses of bean leaves to chronic oxone treatment. Atmos. Environ., 9, 749-53.
                                                                              45,120

RUNECKLES, V.C. and ROSEN, P.M. (1974). Effects of pretreatment with low ozone concentrations on ozone injury to bean and mint. Can. J. Bot., 52, 2607-10.   41,116

RUSH, D.W. and EPSTEIN, E. (1976). Genotypic responses to salinity. Differences between salt-sensitive and salt-tolerant genotypes of the tomato. Pl. Physiol., 57, 162-6.   94,146

RYDER, E.J. (1973). Selecting and breeding plants for increased resistance to air pollutants. In: Naegele, J.A., Ed., Air Pollution Damage to Vegetation, Adv. Chem. Ser. 122, pp. 75-85, Am. Chem. Soc., Washington, D.C.   112

SAUNDERS, P.J.W. (1966). The toxicity of sulphur dioxide to Diplocarpon rosae Wolf causing blackspot of roses. Ann. Appl. Biol., 58, 103-14. 28,49

SAUNDERS, P.J.W. (1976). The estimation of pollution damage. Manchester Univ. Press, 126 pp.   6

SCHRAMEL, M. (1975). Influence of sulfur dioxide on stomatal apertures and diffusive resistance of leaves in various species of cultivated plants under optimum soil moisture and drought conditions. Bull. L'Acad. Pol. Sci. Ser. Sci. Biol., 23, 57-63.   38,121

SCHROEDER, H.A. and BALASSA, J.J. (1963). Cadmium: uptake by vegetables from superphosphate in soil. Science, 140, 819-20.   66

SCHUCK, E.A. (1973). Chemical basis of the air pollution problem. In: Naegele, J.A., Ed., Air Pollution Damage to Vegetation, Adv. Chem. Ser. 122, pp. 1-8, Am. Chem. Soc., Washington, D.C.   8

SEEM, R.C., COLE, H., JR. and LACASSE, N.L. (1972). Suppression of ozone injury to Phaseolus vulgaris L. 'Pinto III' with triarimol and its monochlorophenyl cyclohexyl analogue. Pl. Dis. Rep., 56, 386-90.   43,119

SEEM, R.C., COLE, H., JR, and LACASSE, N.L. (1973). Suppression of ozone injury to Phaseolus vulgaris L. with thiophanate ethyl and its methyl analogue. J. Environ. Qual., 2, 266-8.   43,119

SEIDMAN, G., HINDAWI, I.J. and HECK, W.W. (1965). Environmental conditions affecting the use of plants as indicators of air pollution. J. Air Pollut. Contr. Ass., 15, 168-70.   37,45,117,179

SETTERSTROM, C. and ZIMMERMAN, P.W. (1939). Factors influencing susceptibility of plants to sulfur dioxide injury. Contrib. Boyce Thompson Inst., 10, 155-81.   38

SHARMA, D.P., FERREE, D.C. and HARTMAN, F.O. (1977a). Effect of some soil-applied herbicides on net photosynthesis and growth of apple trees. HortScience, 12, 153-4.   155

SHARMA, D.P., FERREE, D.C. and HARTMAN, F.O. (1977b). Multiple applications of dicofol and dodine sprays on net photosynthesis of apple leaves. HortScience, 12, 154-5.   154

SHAULIS, N.J., KENDER, W.J., PRATT, C. and SINCLAIR, W.A. (1972). Evidence for injury by ozone in New York vineyards. HortScience, 7, 570-2.   160

SHENFELD, L. (1975). Air monitoring. HortScience, 10, 491-2.   50

SHERWOOD, C.H. and ROLPH, G.D. (1970). Ozone protects plants from air pollution with 2,4-D. HortScience, 5, 190 (Abs.)   26,144

SHERWOOD, C.H., WEIGLE, J.L. and DENISEN, E.L. (1970). 2,4-D as an air pollutant: effects on growth of representative horticultural plants. HortScience, 5, 211-3.   86,87,88,89

SHOWMAN, R.E. (1972). Residual effects of sulfur dioxide on the net photosynthesis and respiratory rates of lichen thalli and cultured lichen symbionts. Bryologist, 75, 335-41.   104

SIEGEL, S.M. (1962). Protection of plants against airborne oxidants: cucumber seedlings at extreme ozone levels. Pl. Physiol., 37, 261-6.   42,45,127

SIJ, J.W. and SWANSON, C.A. (1974). Short-term kinetic studies on the inhibition of photosynthesis by sulfur dioxide. J. Environ. Qual., 3, 103-7.   104

SILVIUS, J.E., INGLE, M. and BAER, C.H. (1975). Sulfur dioxide inhibition of photosynthesis in isolated spinach chloroplasts. Pl. Physiol., 56, 434-7.   104

SILVIUS, J.E., BAER, C.H., DODRILL, S. and PATRICK, H. (1976). Photoreduction of sulfur dioxide by spinach leaves and isolated spinach chloroplasts. Pl. Physiol., 57, 799-801.   104

SINCLAIR, W.A., STONE, E.L. and SCHEER, C.F., JR. (1975). Toxicity to hemlocks grown in arsenic-contaminated soil previously used for potato production. HortScience, 10, 35-6.   66,135,193

SINGER, L., ARMSTRONG, W.D. and VATASSERY, G.T. (1967). Fluoride in commercial tea and related plants. Econ. Bot., 21, 285-7.   162

SKELLY, J.M. and LAMBE, R.C. (1974). Diagnosis of air pollution injury to plants. Exten. Div., Virginia State Univ., Publ. 568, 14 pp.   4

SMITH, M.E. (1968). The influence of atmospheric dispersion on the exposure of plants to airborne pollutants. Phytopathology, 58, 1085-8.   6

SMITH, W.H. (1973). Metal contamination of urban woody plants. Environ. Sci. Technol., 7, 631-6.   192

SMITH, W.H. and DOCHINGER, L.S., Eds. (1975). Air Pollution and Metropolitan Woody Vegetation. A Problem Analysis for Environmental Forestry Research. 74 pp. The Pinchot Institute of Environmental Forestry Research, Upper Darby, Pa., U.S.A.   197

SMITH, W.H. and PARKER, J.C. (1966). Prevention of ethylene injury to carnations by low concentrations of carbon dioxide. Nature, 211, 100-1.   166

SOLBERG, R.A. and ADAMS, D.F. (1956). Histological responses of some
plant leaves to hydrogen fluoride and sulfur dioxide. Am. J. Bot.,
43, 755-60.   109

SOLIMAN, M.H. (1976). pH-dependent heterosis of heavy metal-tolerant and
non-tolerant hybrid of the monkey flower, Mimulus guttatus. Nature,
262, 49-51   71

SPALDING, D.H. (1966). Appearance and decay of strawberries, peaches, and
lettuce treated with ozone. ARS, USDA, Mktg. Res. Rept. 756.   11

SPALDING, D.H. (1968). Effects of ozone atmospheres on spoilage of fruits
and vegetables after harvest. ARS,USDA, Mktg. Res. Rept. 801.   11

SPALENY, J. KUTACEK, M. and OPLISTILOVA, K. (1965). On the metabolism of
S^{35} O_2 in the leaves of cauliflower Brassica oleracea var. botrytis L.
Int. J. Air Wat. Pollut., 9, 525-30.   107

SPEDDING, D.T. and THOMAS, W.J. (1973). Effect of sulphur dioxide on the
metabolism of glycolic acid by barley (Hordeum vulgare) leaves. Aust.
J. Biol. Sci., 26, 281-6.   107

SPIERINGS, F. (1964). Differences in susceptibility to damage by HF
between tulip varieties. Toxicology of Fluorine Symp. (1962), pp.
158-61, Schwabe and Co., Verlag, Stuttgart.   180

SPIERINGS, F.H.F.G. (1967a). Chronic discoloration of leaf tips of
gladiolus and its relation to the hydrogen fluoride content of the
air and the fluorine content of the leaves. Neth. J. Pl. Path., 73,
25-8.   174

SPIERINGS, F. (1967b). Method for determining the susceptibility of trees
to air pollution by artificial fumigation. Atmos. Environ., 1, 205-10.
                                                                           53,150

SPIERINGS, F.H.F.G. (1969a). A special type of leaf injury caused by
hydrogen fluoride fumigation of narcissus and nerine. Proc. 1st.
Eur. Congr. Infl. Air Pollut. Plants Animals, Wageningen (1968), 78-89.
                                                                          172,183

SPIERINGS, F. (1969b). Injury to cut flowers of gladiolus by fluoridated
water. Neth. J. Pl. Path., 75, 281-6.   174

SPIERINGS, F.H.F.G. (1971). Influence of fumigations with NO_2 on growth
and yield of tomato plants. Neth. J. Pl. Path., 77, 194-200.   13,144

SRIVASTAVA, H.S., JOLLIFFE, P.A. and RUNECKLES, V.C. (1975a). The
influence of nitrogen supply during growth on the inhibition of gas
exchange and visible damage to leaves by NO_2. Environ. Pollut.,
9, 35-47.   40,103,105,120

SRIVASTAVA, H.S., JOLLIFFE, P.A. and RUNECKLES, V.C. (1975b). The effects
of environmental conditions on the inhibition of leaf gas exchange by
NO_2. Can. J. Bot., 53, 475-82.   103

SRIVASTAVA, H.S., JOLLIFFE, P.A. and RUNECKLES, V.C. (1975c). Inhibition
of gas exchange in bean leaves by NO_2. Can. J. Bot., 53, 466-74.   103

STARKEY, T.E., DAVIS, D.D. and MERRILL, W. (1976). Symptomatology and
susceptibility of ten bean varieties exposed to peroxyacetyl nitrate
(PAN). Pl. Dis. Rep., 60, 480-3.   120,121

STEWART, D.K.R. and ROSS, R.G. (1967). Mercury residues in apples in relation to spray date, variety and chemical composition of fungicide. Can. J. Pl. Sci., 47, 169–74.   73

STOLZY, L.H., TAYLOR, O.C., DUGGER, W.M., JR. and MERSEREAU, J.D. (1964). Physiological changes in and ozone susceptibility of the tomato plant after short periods of inadequate oxygen diffusion to the roots. Proc. Soil Sci. Soc. Am., 28, 305–8.   143

STOLZY, L.H., TAYLOR, O.C., LETEY, J. and SZUSZKIEWICZ, T.E. (1961). Influence of soil-oxygen diffusion rates on susceptibility of tomato plants to air-borne oxidants. Soil Sci., 91, 151–5.   38,143

STRINGER, A., PICKARD, J.A. and LYONS, C.H. (1975). Accumulation and distribution of pp^1-DDT and related compounds in an apple orchard. II. Residues in trees and herbage. Pestic. Sci., 6, 223–32.   154

SUCOFF, E. and HONG, S.G. (1976). Effects of NaCl on cold hardiness of Malus spp. and Syringa vulgaris. Can. J. Bot., 54, 2816–9.   97,98

SUCOFF, E., HONG, S.G. and WOOD, A. (1976). NaCl and twig dieback along highways and cold hardiness of highway versus garden twigs. Can. J. Bot., 54, 2268–74.   97

SULZBACH, C.W. and PACK, M.R. (1972). Effects of fluoride on pollen germination, pollen tube growth, and fruit development in tomato and cucumber. Phytopathology, 62, 1247–53.   145

SUTTON, R. and TING, I.P. (1977). Evidence for repair of ozone induced membrane injury: alteration in sugar uptake. Atmos. Environ., 11, 273–5.   100

TANAKA, H., TAKANASHI, T. and YATAZAWA, M. (1972a). Experimental studies on sulfur dioxide injuries in higher plants. I. Formation of glyoxylate-bisulfite in plant leaves exposed to sulfur dioxide. Wat., Air, Soil Pollut., 1, 205–11.   107

TANAKA, H. TAKANASHI, T., KADOTA, M. and YATAZAWA, M. (1972b). Experimental studies on sulfur dioxide injuries in higher plants. II. Disturbance of amino acid metabolism in plants exposed to sulfur dioxide. Wat., Air, Soil Pollut., 1, 343–6.   107

TANAKA, H., TAKANASHI, T. and YATAZAWA, M. (1974). Experimental studies on SO_2 injuries in higher plants. III. Inhibitory effects of sulfite ion on $^{14}CO_2$ fixation. Wat., Air, Soil Pollut., 3, 11–16.   107

TAYLOR, G.E., JR. and MURDY, W.H. (1975). Population differentiation of an annual plant species, Geranium carolinianum, in response to sulfur dioxide. Bot. Gaz., 136, 212–5.   113

TAYLOR, O.C. (1958). Air pollution with relation to agronomic crops: IV. Plant growth suppressed by exposure to air-borne oxidants (smog). Agron. J., 50, 556–8.   139,156,176

TAYLOR, O.C. (1969). Importance of peroxyacetyl nitrate (PAN) as a phyto-toxic air pollutant. J. Air Pollut. Contr. Ass., 19, 347–51.   12

TAYLOR, O.C. (1973). Acute responses of plants to aerial pollutants. In: Naegele, J.A., Ed., Air Pollution Damage to Vegetation, Adv. Chem. Ser. 122, pp. 9–20, Am. Chem. Soc., Washington, D.C.   11,35

TAYLOR, O.C. (1974). Air pollutant effects influenced by plant-environmental interaction. In: Dugger, M., Ed., Air Pollution Effects on Plant Growth, ACS Symp. Ser. 3, pp. 1-7, Am. Chem. Soc., Washington, D.C.   32

TAYLOR, O.C. (1975). Air pollutant injury to plant processes. HortScience, 10, 501-4.   100

TAYLOR, O.C. and EATON, F.M. (1966). Suppression of plant growth by nitrogen dioxide. Pl. Physiol., 41, 132-5.   13,144

TAYLOR, O.C. and MACLEAN, D.C. (1970). Nitrogen oxides and the peroxyacyl nitrates. In: Jacobson, J.S. and Hill, A.C., Eds., Recognition of Air Pollution Injury to Vegetation: A Pictorial Atlas, pp. E1-E14. Air Pollut. Contr. Ass., Pittsburgh, Pa., U.S.A.   12

TAYLOR, O.C., CARDIFF, E.A. and MERSEREAU, J.D. (1965). Apparent photosynthesis as a measure of air pollution damage. J. Air Pollut. Contr. Ass., 15, 171-3.   64

TAYLOR, O.C., CARDIFF, E.A., MERSEREAU, J.D. and MIDDLETON, J.T. (1958). Effect of air-borne reaction products of ozone and 1-N-hexene vapor (synthetic smog) on growth of avocado seedlings. Proc. Am. Soc. Hort. Sci., 71, 320-5.   155

TAYLOR, O.C., DUGGER, W.M., CARDIFF, E.A. and DARLEY, E.F. (1961). Interactions of light and atmospheric photochemical products (smog) within plants. Nature, 192, 814-6.   34,120

TAYLOR, O.C., STEPHENS, E.R., DARLEY, E.F. and CARDIFF, E.A. (1960). Effect of air-borne oxidants on leaves of pinto bean and petunia. Proc. Am. Soc. Hort. Sci., 75, 435-44.   35

TAYLOR, O.C., THOMPSON, C.R., TINGEY, D.T. and REINERT, R.A. (1975). Oxides of nitrogen. In: Mudd, J.B. and Kozlowski, T.T., Eds., Responses of Plants to Air Pollution, pp. 121-39, Academic Press, Inc., New York.   12

TEMPLE, P.J. (1972). Dose-response of urban trees to sulfur dioxide. J. Air Pollut. Contr. Ass., 22, 271-4.   192

TEMPLE, P.J., LINZON, S.N. and CHAI, B.L. (1977). Contamination of vegetation and soil by arsenic emissions from secondary lead smelters. Environ. Pollut., 12, 311-20.   66

TEN HOUTEN, J.G. (1973). Chemicals, environmental pollution, animals and plants. Meded. Fak. Landb. Wet. Gent, 38, 591-612.   5

TEN HOUTEN, J.G. (1974). Air pollution and horticulture. XIX Int. Hort. Congr. Proc., pp. 57-71.   12,140,167

TER HAAR, G.L., DEDOLPH, R.R., HOLTZMAN, R.B. and LUCAS, H.F., JR. (1969). The lead uptake by perennial ryegrass and radishes from air, water and soil. Environ. Res., 2, 267-71.   71,137

THOMPSON, C.R. (1969). Effects of air pollutants in the Los Angeles basin on citrus. Proc. Ist Int. Citrus Symp., 2, 705-9.   156

THOMPSON, C.R. and KATS, G. (1975). Effects of ambient concentrations of peroxyacetyl nitrate on navel orange trees. Environ. Sci. Technol., 9, 35-8.   156

THOMPSON, C.R. and TAYLOR, O.C. (1967). Reduction of fruit drop by navel
oranges with antioxidant dusts and girdling. HortScience, 2, 103-4. 42,156

THOMPSON, C.R. and TAYLOR, O.C. (1969). Effects of air pollutants on
growth, leaf drop, fruit drop, and yield of citrus trees. Environ.
Sci. Technol., 3, 934-40.   156

THOMPSON, C.R., HENSEL, E. and KATS, G. (1969). Effects of photochemical
air pollutants on Zinfandel grapes. HortScience, 4, 222-4.   160

THOMPSON, C.R., HENSEL, E.G. and KATS, G. (1973). Outdoor-indoor levels
of six air pollutants. J. Air Pollut. Contr. Ass., 23, 881-6.   20

THOMPSON, C.R., KATS, G. and CAMERON, J.W. (1976). Effects of ambient
photochemical air pollutants on growth, yield, and ear characters of
two sweet corn hybrids. J. Environ. Qual., 5, 410-2.   125

THOMPSON, C.R., KATS, G. and HENSEL, E. (1972). Effects of ambient levels
of ozone on navel oranges. Environ. Sci. Technol., 6, 1014-6.   156,157

THOMPSON, C.R., HENSEL, E.G., KATS, G. and TAYLOR, O.C. (1970). Effects
of continuous exposure of navel oranges to nitrogen dioxide. Atmos.
Environ., 4, 349-55.   157

THOMPSON, C.R., TAYLOR, O.C., THOMAS, M.D. and IVIE, J.O. (1967). Effects
of air pollutants on apparent photosynthesis and water use by citrus
trees. Environ. Sci. Technol., 1, 644-50.   156

THOMPSON, N.P. and BROOKS, R.F. (1976). Disappearance of dislodgable
residues of five organophosphate pesticides on citrus leaves and fruit
during dry and wet weather in Florida. Arch. Environ. Contam. Toxicol.,
5, 55-61.   80

THOMSON, W.W. (1975). Effects of air pollutants on plant ultrastructure.
In: Mudd, J.B. and Kozlowski, T.T., Eds., Responses of Plants to Air
Pollution, pp. 179-94, Academic Press, Inc., New York.   109

THOMSON, W.W., DUGGER, W.M., JR. and PALMER, R.L. (1966). Effects of ozone
on the fine structure of the palisade parenchyma cells of bean leaves.
Can. J. Bot., 44, 1677-82.   109

THOMSON, W.W., NAGAHASHI, J. and PLATT, K. (1974). Further observations
on the effects of ozone on the ultrastructure of leaf tissue. In:
Dugger, M., Ed., Air Pollution Effects on Plant Growth. Am. Chem. Soc.
Symp. Ser. 3, pp. 83-93, Am. Chem. Soc., Washington, D.C.   109

THORNE, L. and HANSON, G.P. (1972). Species differences in rates of vege-
tal ozone absorption. Environ. Pollut., 3, 303-12.   32

THORNE, L. and HANSON, G.P. (1976). Relationship between genetically
controlled ozone sensitivity and gas exchange rate in Petunia hybrida
Vilm. J. Am. Soc. Hort. Sci., 101, 60-3.   113

TING, I.P., PERCHOROWICZ, J. and EVANS, L. (1974). Effect of ozone on
plant cell membrane permeability. In: Dugger, M., Ed., Air Pollution
Effects on Plant Growth, Am. Chem. Soc. Symp. Ser. 3, pp. 8-21, Am.
Chem. Soc., Washington, D.C.   100

238

TINGEY, D.T. (1974). Ozone induced alterations in the metabolite pools and enzyme activities of plants. In: Dugger, M., Ed., Air Pollution Effects on Plant Growth, Am. Chem. Soc. Symp. Ser. 3, pp. 40-57, Am. Chem. Soc., Washington, D.C.   35,106

TINGEY, D.T. and REINERT, R.A. (1975). The effect of ozone and sulphur dioxide singly and in combination on plant growth. Environ. Pollut., 9, 117-25.   137

TINGEY, D.T., DUNNING, J.A. and JIVIDEN, G.M. (1973a). Radish root growth reduced by acute ozone exposures. Proc. 3rd Int. Clean Air Congr., A154-6. VDI-Verlag GmbH, Dusseldorf.   136

TINGEY, D.T., FITES, R.C. and WICKLIFF, C. (1973b). Foliar sensitivity of soybeans to ozone as related to several leaf parameters. Environ. Pollut., 4, 183-92.   11

TINGEY, D.T., FITES, R.C. and WICKLIFF, C. (1973c). Ozone alteration of nitrate reduction in soybean. Physiol. Pl., 29, 34-8.   101

TINGEY, D.T., HECK, W.W. and REINERT, R.A. (1971a). Effect of low concentrations of ozone and sulfur dioxide on foliage, growth and yield of radish. J. Am. Soc. Hort. Sci., 96, 369-71.   136,137

TINGEY, D.T., REINERT, R.A., DUNNING, J.A. and HECK, W.W. (1971b). Vegetation injury from the interaction of nitrogen dioxide and sulfur dioxide. Phytopathology, 61, 1506-11.   23,24

TINGEY, D.T., REINERT, R.A., DUNNING, J.A. and HECK, W.W. (1973d). Foliar injury responses of eleven plant species to ozone/sulfur dioxide mixtures. Atmos. Environ., 7, 201-8.   24,123

TINGEY, D.T., STANDLEY, C. and FIELD, R.W. (1976). Stress ethylene evolution: a measure of ozone effect on plants. Atmos. Environ., 10, 969-74.   11,52

TJIA, B.O.S., ROGERS, M.N. and HARTLEY, D.E. (1969). Effects of ethylene on morphology and flowering of Chrysanthemum morifolium Ramat. J. Am. Soc. Hort. Sci., 94, 35-9.   169

TODD, G.W. (1958). Effect of ozone and ozonated 1-hexene on respiration and photosynthesis of leaves. Pl. Physiol., 33, 416-20.   102,114

TODD, G.W. and ARNOLD, W.N. (1961). An evaluation of methods used to determine injury to plant leaves by air pollutants. Bot. Gaz., 123, 151-4.   49

TODD, G.W. and GARBER, M.J. (1958). Some effects of air pollutants on the growth and productivity of plants. Bot. Gaz., 120, 75-80.   114,130

TOMLINSON, H. and RICH, S. (1967). Metabolic changes in free amino acids of bean leaves exposed to ozone. Phytopathology, 57, 972-4.   105

TOMLINSON, H. and RICH, S. (1970a). Lipid peroxidation, a result of injury in bean leaves exposed to ozone. Phytopathology, 60, 1531-2.   105

TOMLINSON, H. and RICH, S. (1970b). Disulfides in bean leaves exposed to ozone. Phytopathology, 60, 1842-3.   106

TOMLINSON, H. and RICH. S. (1971). Effect of ozone on sterols and sterol derivatives in bean leaves. Phytopathology, 61, 1404-5.   100,106

TOMLINSON, H. and RICH, S. (1973). Anti-senescent compounds reduce injury and steroid changes in ozonated leaves and their chloroplasts. Phytopathology, 63, 903-6.   45,46

TOUCHTON, J.T. and BOSWELL, F.C. (1975). Use of sewage sludge as a green-house soil amendment. II. Influence on plant growth and constituents. Agr. Environ., 2, 243-50.   76,146

TOWNSEND, A.M. (1974). Sorption of ozone by nine shade tree species. J. Am. Soc. Hort. Sci., 99, 206-8.   31

TOWNSEND, A.M. and DOCHINGER, L.S. (1974). Relationship of seed source and developmental stage to the ozone tolerance of Acer rubrum seedlings. Atmos. Environ., 8, 957-64.   188

TRAPPE, J.M., STAULY, E.A., BENSON, N.R. and DUFF, D.M. (1973). Mycorrhizal deficiency of apple trees in high arsenic soils. HortScience, 8, 52-3.   66

TRESHOW, M. (1968). The impact of air pollutants on plant populations. Phytopathology, 58, 1108-13.   4

TRESHOW, M. (1970a). Environment and Plant Response. McGraw-Hill Book Co., New York.   4

TRESHOW, M. (1970b). Ozone damage to plants. Environ. Pollut., 1, 155-61.   10

TRESHOW, M. (1971). Fluorides as air pollutants affecting plants. Annu. Rev. Phytopath., 9, 21-44.   16

TRESHOW, M. (1975). Interaction of air pollutants and plant diseases. In: Mudd, J.B. and Kozlowski, T.T., Eds., Responses of Plants to Air Pollution, pp. 307-33, Academic Press, Inc., New York.   27

TRESHOW, M. and HARNER, F.M. (1968). Growth responses of Pinto bean and alfalfa to sublethal fluoride concentrations. Can. J. Bot., 46, 1207-10.   121

TRESHOW, M. and PACK, A.C. (1970). Fluoride. In: Jacobson, J.S. and Hill, A.C., Eds., Recognition of Air Pollution Injury to Vegetation: A Pictorial Atlas, pp. D1-D17, Air Pollut. Contr. Ass., Pittsburgh, Pa.   16,17

TRESHOW, M., DEAN, G. and HARNER, F.M. (1967). Stimulation of tobacco mosaic virus-induced lesions on bean by fluoride. Phytopathology, 57, 756-8.   28

TURNER, M.A. (1973). Effect of cadmium treatment on cadmium and zinc uptake by selected vegetable species. J. Environ. Qual., 2, 118-9.   67

TURNER, N.C., RICH, S. and WAGGONER, P.E. (1973). Removal of ozone by soil. J. Environ. Qual., 2, 259-64.   31

TWEEDY, J.A. and RIES, S.K. (1966). Fruit tree tolerance to two triazines. Weeds, 14, 268-9.   86

U.S. DEPT. AGRIC. (1973). Air Pollution Damages Trees. Forest Service, Upper Darby, Pa., 32 pp.   187

UNSWORTH, M.H., BISCOE, P.V. and BLACK, V. (1976). Analysis of gas
exchange between plants and polluted atmospheres. In: Mansfield,
T.A., Ed., Effects of Air Pollutants on Plants, Soc. Exptl. Biol.
Sem. Ser. 1, pp. 6-16, Cambridge Univ. Press, Cambridge.   30

UNSWORTH, M.H., BISCOE, P.V. and PINCKNEY, H.R. (1972). Stomatal responses
to sulphur dioxide. Nature, 239, 458-9.   38

UOTA, M. (1969). Carbon dioxide suppression of ethylene-induced sleepiness of carnation blooms. J. Am. Soc. Hort. Sci., 94, 598-601.   166

VANDEGRIFT, A.E., SHANNON, L.J., SALLEE, E.E., GORMAN, P.G. and PARK, W.R.
(1971). Particulate air pollution in the United States. J. Air Pollut.
Contr. Ass., 21, 321-8.   20

VAN HAUT, H. and STRATMANN, H. (1970). Color-Plate Atlas of the Effects
of Sulfur Dioxide on Plants. Verlag W. Girardet, Essen. 206 pp.   13,15

VAN RAAY, A. (1969). The use of indicator plants to estimate air pollution
by SO$_2$ and HF. Proc. 1st Eur. Congr. Infl. Air Pollut. Plants Animals,
Wageningen (1968), 319-28.   50

VERKROOST, M. (1974). The effect of ozone on photosynthesis and respiration of Scenedesmus obtusiusculus Chod, with a general discussion of
effects of air pollutants in plants. Meded. Landbhogesch. Wageningen,
74, 1-78.   30,100,103,107

VITOSH, M.L. and CHASE, R.W. (1973). Speckle leaf of potato as affected
by fertilizer and water management. Am. Potato J., 50, 311-4.
                                                                  37,39,134,135

WAGGONER, P.E. (1971). Plants and polluted air. Bioscience, 21, 455-9. 4,11

WAGGONER, P.E. (1972). Role of plants in improving the environment. J.
Environ. Qual., 1, 123-7.   31

WALKER, K.C. (1970). The effects of horticultural practices on man and
his environment. HortScience, 5, 239-42.   6

WALKER, J.T. and BARLOW, J.C. (1974). Response of indicator plants to
ozone levels in Georgia. Phytopathology, 64, 1122-7.   44,52,177

WALKER, J.T., HIBBEN, C.R. and ALLISON, J.C. (1975). Cultivar ratings for
susceptibility and resistance to the leaf roll-necrosis disorder of
lilac. J. Am. Soc. Hort. Sci., 100, 627-30.   190,191

WALLACE, A., SOUFI, S.M., CHA, J.W. and ROMNEY, E.M. (1976). Some effects
of chromium toxicity on bush bean plants grown in soil. Pl. Soil, 44,
471-3.   70,123

WALSH, L.M., ERHARDT, W.H. and SEIBEL, H.D. (1972). Copper toxicity in
snapbeans (Phaseolus vulgaris L.). J. Environ. Qual., 1, 197-200.   70,122

WANDER, I.W. and MCBRIDE, J.J., JR. (1956). Chlorosis produced by fluorine
on citrus in Florida. Science, 123, 933-4.   15

WEHRMANN, J. (1974). Polluted water problems in horticulture. Proc. XIX
Int. Hort. Congr., 73-82.   91,92

WEI, L.-L. and MILLER, G.W. (1972). Effect of HF on the fine structure
of mesophyll cells from Glycine max Merr. Fluoride, 5, 67-73.   111

WEIGLE, J.L., DENISEN, E.L. and SHERWOOD, C.H. (1970). 2,4-D as an air
pollutant: effects on market quality of several horticultural crops.
HortScience, 5, 213-4.   86

WEINSTEIN, L.H. (1961). Effects of atmospheric fluoride on metabolic
constituents of tomato and bean leaves. Contrib. Boyce Thompson
Inst., 21, 215-31.   108,122,145

WEINSTEIN, L.H. (1977). Fluoride and plant life. J. Occup. Med., 19,
49-78.   16

WEINSTEIN, L.H. and MCCUNE, D.C. (1970). Field surveys, vegetation
sampling, and air and vegetation monitoring. In: Jacobson, J.S. and
Hill, A.C., Eds., Recognition of Air Pollution Injury to Vegetation:
A Pictorial Atlas, pp. G1-G4, Air Pollut. Contr. Ass., Pittsburgh, Pa.   49

WEINSTEIN, L.H., MCCUNE, D.C., ALUISIO, A.L. and VAN LEUKEN, P. (1975).
The effect of sulphur dioxide on the incidence and severity of bean
rust and early blight of tomato. Environ. Pollut., 9, 145-55.   28,48

WELLBURN, A.R., MAJERNIK, O. and WELLBURN, F.A.M. (1972). Effects of SO_2
and NO_2 polluted air upon the ultrastructure of chloroplasts. Environ.
Pollut., 3, 37-49.   111

WELLBURN, A.R., CAPRON, T.M., CHAN, H.-S. and HORSMAN, D.C. (1976). Bio-
chemical effects of atmospheric pollutants on plants. In: Mansfield,
T.A., Ed., Effects of Air Pollution on Plants. Soc. Exptl. Biol.,
Sem. Ser. I, pp. 105-14, Cambridge Univ. Press, Cambridge.   101

WESTING, A.H. (1969). Plants and salt in the roadside environment.
Phytopathology, 59, 1174-81.   95,97

WILL, J.B. and SKELLY, J.M. (1974). The use of fertilizer to alleviate
air pollution damage to white pine (Pinus strobus) Christmas trees.
Pl. Dis. Rep., 58, 150-4.   189

WILLIAMS, C.H. and DAVID, D.J. (1973). The effect of superphosphate on
the cadmium content of soils and plants. Aust. J. Soil Res., 11,
43-56.   66

WILLIAMS, D.J. and MOSER, B.C. (1975). Critical level of airborne sea
salt inducing foliar injury to bean. HortScience, 10, 615-6.   123

WILLIAMS, J.H., KINGHAM, H.G., COOPER, B.J. and EAGLE, D.J. (1977).
Growth regulator injury to tomatoes in Essex, England. Environ. Pollut.,
12, 149-57.   84

WILLIX, R. (1976). An introduction to the chemistry of atmospheric pollu-
tants. In: Mansfield, T.A., Ed., Effects of Air Pollutants on
Plants, Soc. Exptl. Biol. Sem. Ser. I, pp. 161-84, Cambridge Univ.
Press, Cambridge.   8

WILTON, A.C., MURRAY, J.J., HEGGESTAD, H.E. and JUSKA, F.V. (1972).
Tolerance and susceptibility of Kentucky bluegrass (Poa pratensis L.)
cultivars to air pollution: in the field and in an ozone chamber.
J. Environ. Qual., 1, 112-4.   185

WOLTING, H.G. (1971). Gevoeligheid van tulpen voor HF bij langdurige
begassing met zeer lage concentraties. Bedrijfsontwikkeling 2, 53-7.   180

WOLTING, H.G. (1975). Synergism of hydrogen fluoride and leaf necrosis
on freesias. Neth. J. Pl. Path., 81, 71-7.   173

WOOD, F.A. (1968). Sources of plant-pathogenic air pollutants. Phytopathology, 58, 1075-84.   4,8

WOOD, F.A. and DRUMMOND, D.B. (1974). Response of eight cultivars of
chrysanthemum to peroxyacetyl nitrate. Phytopathology, 64, 897-8.   169

WOOD, T. and BORMANN, F.H. (1974). The effects of an artificial acid mist
upon the growth of Betula alleghaniensis Britt. Environ. Pollut., 7,
259-68.   21

WOOD, T. and BORMANN, F.H. (1977). Short-term effects of a simulated acid
rain upon the growth and nutrient relations of Pinus strobus L.
Wat., Air, Soil Pollut., 7, 479-88.   22

WRIGHT, M.J., Ed. (1976). Plant adaptation to mineral stress in problem
soils. Cornell Univ. Agr. Exp. Sta. Spec. Publ., 420 pp.   65

WUKASCH, R.T. and HOFSTRA, G. (1977a). Ozone and Botrytis spp. interaction
in onion leaf dieback: field studies. J. Am. Soc. Hort. Sci., 102,
543-6.   28,129

WUKASCH, R.T. and HOFSTRA, G. (1977b). Ozone and Botrytis interactions
in onion-leaf dieback: open-top chamber studies. Phytopathology,
67,1080-4.   28,129

YARON, B., ZIESLIN, N. and HALEVY, A.H. (1969). Response of Baccara roses
to saline irrigation. J. Am. Soc. Hort. Sci., 94, 481-4.   194

YARWOOD, C.E. and MIDDLETON, J.T. (1954). Smog injury and rust infection.
Pl. Physiol. 29, 393-5.   27

YEE-MEILER, D. (1974). Uber den einfluss fluorhaltiger fabrikabgase auf
den phenolgehalt von fichtennadein. Eur. J. Pl. Path., 4, 214-21.   108

ZEEVAART, A.J. (1976). Some effects of fumigating plants for short periods
with NO_2. Environ. Pollut., 11, 97-108.   13,35,48,105,131

ZIEGLER, I. (1972). The effect of SO_3^{--} on the activity of ribulose-1,
5-diphosphate carboxylase in isolated spinach chloroplasts. Planta,
103, 155-63.   101

ZIEGLER, I. (1973a). Effect of sulphite on phosphoenolpyruvate carboxylase
and malate formation in extracts of Zea mays. Phytochemistry, 12,
1027-30.   101

ZIEGLER, I. (1973b). The effect of air-polluting gases on plant metabolism.
In: Coulson, F. and Korte, F., Eds., Environmental Quality and Safety.
Vol. II. Global Aspects of Chemistry, Toxicology and Technology as
Applied to the Environment. Academic Press, New York.   30,100,103,107

ZEIGLER, I. (1974a). Action of sulphite on plant malate dehydrogenase.
Phytochemistry, 13, 2411-16.   101

ZIEGLER, I. (1974b). Malate dehydrogenase in _Zea_ _mays_: properties and inhibition by sulfite. Biochem. Biophys. Acta, 364, 28-37.   101

ZIMDAHL, R.L. (1976). Entry and movement in vegetation of lead derived from air and soil sources. J. Air Pollut. Contr. Ass., 26, 655-60.   72

ZIMMERMAN, P.W. and CROCKER, W. (1934). Toxicity of air containing sulphur dioxide gas. Contrib. Boyce Thompson Inst., 6, 455-70.   13,33,38

ZIMMERMAN, P.W. and HITCHCOCK, A.E. (1956). Susceptibility of plants to hydrofluoric acid and sulfur dioxide gases. Contrib. Boyce Thompson Inst., 18, 263-79.   15,16,17

ZIMMERMAN, P.W., HITCHCOCK, A.E. and CROCKER, W. (1931). The effect of ethylene and illuminating gas on roses. Contrib. Boyce Thompson Inst., 3, 459-81.   192

APPENDIX I

GLOSSARY OF SELECTED AIR POLLUTION TERMS

From Phytopathology News, 8(8), 1974. Selected terms from a glossary prepared by the Pollution Damage on Plants Committee of the American Phytopathological Society.

acid aerosol -- acid droplets, generally less than 1 micron diam, suspended in air.

acid rain -- rain that contains as principal components the hydrolyzed end products from oxidized sulfur or nitrogen and halogen compounds.

acute injury -- injury, usually involving necrosis, which develops within several hours to a few days after a short term exposure to a pollutant, and expressed as fleck, scorch, bifacial necrosis, etc.

additive effects -- the combined effects of more than one pollutant acting simultaneously or in succession to give a total plant response equal to the sum of the independent effects.

aerosol -- a colloidal system in which solid or liquid particles, usually less than 1 micron diam, are suspended in a gas, e.g., smoke, some fogs.

air pollution episode -- a period of sustained high levels of atmospheric pollution often coinciding with periods of thermal inversions and low wind speeds.

air quality standards -- air pollutant concentrations which cannot legally be exceeded during fixed time intervals within specified geographical areas.

ambient air -- air surrounding a given locus; the outside air.

antagonism -- when the combined effect of two or more pollutants is less than the sum of their independent effects; the antonym of synergism.

antioxidant, antiozonant -- a chemical that detoxifies or decreases the oxidizing effects of oxidants or O_3, e.g. ascorbic acid, 4,4-'dioctyl-diphenylamine.

banding -- a foliar symptom characterized by a limited zone of necrotic or discoloured tissue traversing the leaf, e.g., the band on petunia leaves injured by PAN.

bifacial necrosis -- death of plant tissues extending from the adaxial to the abaxial leaf surface.

bioindicator, biological indicator -- plant species, varieties or cultivars sufficiently sensitive to a specific pollutant to make them useful as indicators for the presence of that pollutant.

bronzing -- a brown discolouration that usually appears on the abaxial surface of leaves, and is often an advanced stage of the silvering or glazing typical of injury by PAN and other oxidants.

chronic injury -- injury which develops only after long-term or repeated exposure to an air pollutant, and expressed as chlorosis, bronzing, premature senescence, reduced growth, etc., can include necrosis.

damage -- a measure of the decrease in economic or aesthetic value resulting from plant injury by pollutants.

dose -- a measured concentration of toxicant for a known duration of time (concentration per unit time) to which a receptor is exposed.

epinasty -- outward and downward curvature of a plant part, usually leaves.

fleck -- white to tan necrotic lesions up to a few millimeters in length or diam, usually confined to the adaxial surface of leaves; a characteristic response of tobacco to O_3.

hidden or "invisible" injury -- plant injury not characterized by overt symptoms, e.g., decreased yields, lower quality of plant products.

injury -- any change in the appearance and/or function of a plant that is deleterious to the plant.

intercostal -- leaf tissue between veins; interveinal.

marginal necrosis -- death of tissue at the periphery of a leaf blade.

monitoring -- the use of gas sensing instruments or other devices to measure the concentrations of pollutants.

mottle -- irregular, diffuse patterns of chlorotic areas interspersed with normal green leaf tissue.

oxidant -- a substance capable of oxidizing a reference substance, that substance itself incapable of being oxidized by atmospheric oxygen; refers to several oxidizing gases in the atmosphere, particularly O_3, NO_2 and peroxyacetyl nitrate (PAN); those compounds capable of liberating iodine from neutral buffered potassium iodide solutions.

particulates -- finely divided particles of solid or liquid matter, e.g., dust, smoke, aerosols.

photochemical smog -- a combination of photochemical oxidants, smoke, fumes and aerosols in the polluted atmosphere.

photochemical reaction -- a chemical reaction dependent upon radiation in the form of light energy.

phytotoxicant -- any agent that becomes toxic to plants.

pollutant (air) -- any gas, liquid or solid air contaminant that causes undesirable effects on living organisms or materials.

predisposition -- the tendency of nongenetic conditions to affect the susceptibilitbility of plants to a later stress or infection.

premature senescence -- an accelerated rate of the normal phenological events, e.g., early maturation of leaves.

primary pollutants -- pollutants which are emitted directly from an identifiable source, e.g., SO_2 from stacks.

secondary pollutants -- pollutants produced in the air by reactions involving primary pollutants and/or other atmospheric constituents; e.g., oxidants produced by photochemical reactions.

silvering, silverleaf -- a symptom of leaves or fleshy tissues caused by the abnormal increase of subepidermal air spaces; often induced by PAN on the abaxial surface of leaves after injury to spongy mesophyll cells.

smog -- (General) a mixture of smoke and fog. London type - a mixture of coal smoke and fog, containing enough SO_2 to impart chemical reducing properties to the mixture. Los Angeles type - a mixture of photochemical oxidants, primary pollutants from petroleum combustion, smoke and aerosols.

stomatal resistance -- the degree to which stomata impede conductance of gases, water vapour or fluids between internal tissues of the leaf and the ambient air.

synergism -- when the combined effect of two or more independent treatments is greater than the sum of each treatment alone.

threshold dose -- the minimum dose of a pollutant necessary to induce plant injury.

tipburn, tip necrosis -- necrosis of apical tissues of leaves; includes only a small percentage of the entire leaf.

tolerance -- the capacity of plants to withstand the effects of pollutants without significant reduction in quality or quantity of yield.

APPENDIX II

SUMMARY OF TYPICAL RECOMMENDED NATIONAL AIR QUALITY OBJECTIVES

		Air contaminant							
		SO_2		CO		Oxidants (O_3)		NO_2	
		($\mu g/m^3$)	(ppm)	(mg/m^3)	(ppm)	($\mu g/m^3$)	(ppm)	($\mu g/m^3$)	(ppm)
Maximum	1 h	900	0.34	35	30	160	0.08	400	0.21
acceptable	8 h			15	13				
limit	24 h	300	0.11			50	0.025	200	0.11
	1 yr	60	0.02			30	0.015	100	0.05
Maximum	1 hr	450	0.17	15	13	100	0.05		
desirable	8 h			6	5				
limit	24 h	150	0.06			30	0.015		
	1 yr	30	0.01			20	0.01	60	0.03

Conversion factors to convert mass units to volume units of concentration (at $25^{\circ}C$ and 760 mm Hg).

SO_2 3.82×10^{-4}

CO   0.87

Oxidants (O_3) 5.1×10^{-4}

NO_2 5.32×10^{-4}

From Fisheries and Environment Canada (1976)

APPENDIX III

INDEX OF COMMON AND SCIENTIFIC PLANT NAMES

Underlined page numbers indicate the section devoted to the species.

Willow – black	Salix nigra Marsh.	195
– Arctic blue	S. purpurea L. var nana Hort.	193
Witch hazel	Hamamelis L. spp.	195
Yew	Taxus L. spp.	195
– dense Anglojap	T. densiformis Hort.	188
– Hatfield Anglojap	T. media hatfieldi Rehd.	188
Zinnia	Zinnia elegans Jacq.	26,183
Zoysia	Zoisia Willd. spp.	184–6

References for Scientific Plant Names: BAILEY, L.H. 1949. Manual of
Cultivated Plants, The MacMillan Company, New York or authors of
journal articles.

APPENDIX IV

INDEX OF CHEMICALS

An alphabetical listing of names of chemicals referred to in the text. In
general, the names are those used in the applicable reference(s).

INDEX OF SUBSECTIONS

Subsections not included in CONTENTS